D0713580

Math for Welders

Fifth Edition

by
Nino Marion
St. Clair College

Publisher
Goodheart-Willcox Company, Inc.
Tinley Park, IL
www.g-w.com

Copyright © 2013
by
The Goodheart-Willcox Company, Inc.

Previous editions copyright 2006, 2001, 1996, 1990

All rights reserved. No part of this work may be reproduced, stored, or
transmitted in any form or by any electronic or mechanical means, including
information storage and retrieval systems, without the prior written permission of
The Goodheart-Willcox Company, Inc.

Manufactured in the United States of America.

Library of Congress Catalog Card Number 2011052760

ISBN 978-1-60525-900-0

5 6 7 8 9 – 13 – 18 17 16

The Goodheart-Willcox Company, Inc. Brand Disclaimer: Brand names, company names, and illustrations for products and services included in this text are provided for educational purposes only and do not represent or imply endorsement or recommendation by the author or the publisher.

The Goodheart-Willcox Company, Inc. Safety Notice: The reader is expressly advised to carefully read, understand, and apply all safety precautions and warnings described in this book or that might also be indicated in undertaking the activities and exercises described herein to minimize risk of personal injury or injury to others. Common sense and good judgment should also be exercised and applied to help avoid all potential hazards. The reader should always refer to the appropriate manufacturer's technical information, directions, and recommendations; then proceed with care to follow specific equipment operating instructions. The reader should understand these notices and cautions are not exhaustive.

The publisher makes no warranty or representation whatsoever, either expressed or implied, including but not limited to equipment, procedures, and applications described or referred to herein, their quality, performance, merchantability, or fitness for a particular purpose. The publisher assumes no responsibility for any changes, errors, or omissions in this book. The publisher specifically disclaims any liability whatsoever, including any direct, indirect, incidental, consequential, special, or exemplary damages resulting, in whole or in part, from the reader's use or reliance upon the information, instructions, procedures, warnings, cautions, applications, or other matter contained in this book. The publisher assumes no responsibility for the activities of the reader.

The Goodheart-Willcox Company, Inc. Internet Disclaimer: The Internet resources and listings in this Goodheart-Willcox Publisher product are provided solely as a convenience to you. These resources and listings were reviewed at the time of publication to provide you with accurate, safe, and appropriate information. Goodheart-Willcox Publisher has no control over the referenced websites and, due to the dynamic nature of the Internet, is not responsible or liable for the content, products, or performance of links to other websites or resources. Goodheart-Willcox Publisher makes no representation, either expressed or implied, regarding the content of these websites, and such references do not constitute an endorsement or recommendation of the information or content presented. It is your responsibility to take all protective measures to guard against inappropriate content, viruses, or other destructive elements.

Library of Congress Cataloging-in-Publication Data

Marion, Nino.
 Math for welders / by Nino Marion. - - 5th ed.
 p. cm.
Includes index.
 ISBN 978-1-60525-900-0
 1. Welding—Mathematics. I. Title.

TS227.2.M37   2013
513'.13024671--dc23

2011052760

Introduction

Math for Welders is a combination textbook and workbook that teaches basic mathematics skills and provides practical exercises useful in the welding field. The textbook covers six areas of instruction including:

- Whole numbers
- Common fractions
- Decimal fractions
- Measurement
- Percentages
- SI metric system

The topics are presented in a step-by-step approach with clear examples. This makes learning easy and also improves your understanding of the basics.

Math for Welders lets you apply your learning by using many drills and exercises. Space is provided in the book to work the problems and to record your answers. By referring to the answers to the odd-numbered practice problems found in the back of the book, you will be able to check your progress as you study. The welding-related problems are designed to sharpen your application of basic mathematics to everyday situations.

An understanding of the material in this book is just as important to your career development as keeping your equipment in top shape or developing the ability to weld in various positions. Learn the skills taught here and become successful on the job!

Nino Marion

About the Author

After graduating from Wayne State University, Mr. Marion invested several years working for Chrysler Canada. He then transitioned to a teaching position at St. Clair College, focusing on math and technical drawing. Mr. Marion used his collective knowledge and experience to author *Math for Welders* as a career capstone.

Copyright by The Goodheart-Willcox Co., Inc.

4

Contents

Section 1
Whole Numbers

Copyright by The Goodheart-Willcox Co., Inc.

Section 2
Common Fractions

Section 3
Decimal Fractions

Copyright by The Goodheart-Willcox Co., Inc.

Section 4
Measurement

Copyright by The Goodheart-Willcox Co., Inc.

Copyright by The Goodheart-Willcox Co., Inc.

Copyright by The Goodheart-Willcox Co., Inc.

Section 1
Whole Numbers

Section Objectives

After studying this section, you will be able to:

- Practice good math work habits
- Explain the Arabic number system
- Demonstrate how to round off whole numbers
- Define what is meant by a denominate number
- Perform addition of whole numbers
- Perform subtraction of whole numbers
- Perform multiplication of whole numbers
- Perform division of whole numbers
- Demonstrate how to check answers for four operations

Copyright by The Goodheart-Willcox Co., Inc.

Unit 1
Introduction to Whole Numbers

Key Terms

approximate numbers   decimal number   denominate numbers   rounding
Arabic number system   system   place value

Introduction

A mastery of mathematics is one of the skills expected of you as a welder. Fortunately, the method of acquiring math skills is no different than any other skill. You learn the principles and then practice them in repeated applications until they become "second nature."

What about the questions: "Do I need to know math if I know how to use an electronic calculator?" "Is it necessary to develop a skill in mental calculation and a mastery of math facts?" There are some practical reasons why the answer to these questions is YES. First, your employer, supervisor, etc., will expect you to be knowledgeable in math. They will assume that, as a tradesperson, basic math skills and facts will be part of the mental skills you bring to the job. If you begin your training in math by relying on a calculator, you will do yourself a great disservice. A calculator will not give you the confidence and understanding that comes with building a skill.

Also, when a group of welders are discussing the math aspects of a job (dimensions, weights, volumes, costs, etc.) you must be able to "keep up" with the discussion. If your math skills are weak, you will likely feel very uncomfortable about joining in on this part of your work. Besides, there is certain admiration granted to a tradesperson by fellow workers who recognize that the person is clearly proficient in math.

Math Work Habits

Although math is almost entirely a mental activity, there is a small but important tangible component to it. All your calculations, dimensions, and notes must be readable. Correct results are, of course, the "final product" of math, but the calculations should be done with a high degree of neatness and care. Orderly work in math is one of the important factors in arriving at correct answers. Listed are some suggestions you might find useful.

- Write your numbers fairly large.

- A poorly written **4** and **9** are easily confused.

- Try writing 7 as 7̸. The number 7, when handwritten, can sometimes be confused with 1 or 2. The slash helps to avoid this confusion.

- Always carry a thick, sharp pencil while in the shop or in the field.

- When dealing with numbers, do not rush. In fact, deliberately slow yourself down. If you consider the amount of time it takes to turn out a job in the shop, there is no measurable gain in time by calculating and writing figures quickly.

Our Number System

Our number system is sometimes called the **Arabic number system**, because this system was developed in Arab culture. It is a system based on ten digits, namely, 0, 1, 2, 3, 4, 5, 6, 7, 8, 9. Our

Copyright by The Goodheart-Willcox Co., Inc.

number system is also often referred to as the **decimal number system**. The word "decimal" is a word derived from Latin meaning "based on ten."

Practically any number can be expressed by arranging these numbers in a certain order. When a series of digits is written, such as 2497, each of the numbers in that group takes on a certain value based on its **place value** in the lineup. In the example of 2,497, the 7 is considered a 7 because it is in the ones (or units) position. The 9 has a value of 90 because it is in the tens position. You can think of it as $9 \times 10 = 90$. The 4, because of its place in the lineup, has a value of 400. It is in the hundreds column, so you can think of it as $4 \times 100 = 400$. The 2 is placed in the thousands position and has a value of 2,000.

Each of the positions in a line of figures has a value and a name. Listed below are some of the names and their place position.

To help make numbers easier to read, a comma is usually placed after every third digit counting from the right. Here is an example. The number 3,894,076,215 is divided by commas and is read as three billion, eight hundred ninety four million, seventy six thousand, two hundred fifteen. The 3 has a place value of three billion. The 8 has a place value of eight hundred million, the 9 has a place value of ninety million, and so on.

Rounding Numbers

There are some occasions in math where extreme accuracy is not required. Cost estimates for most jobs are not usually exact. Precise weights of large weldments are often not necessary. In these and many other situations, figures may be rounded. **Rounding** is an approximating or adjusting of a number by increasing or decreasing a significant digit and cutting off or reducing to zero the least significant digit or digits. Since rounded figures are not perfectly accurate, they are called **approximate numbers**.

How to Round Numbers

Here is a method for rounding numbers:
1. First, find out to what place value you are rounding. Normally, you are asked to "round to the nearest hundred or thousand," etc.
2. Place a little tick mark over the digit in that place position.
3. If the figure to the right of that number is 5 or more, increase the ticked number by one.
4. If the figure to the right is less than five, do not change the ticked number.

Copyright by The Goodheart-Willcox Co., Inc.

5. Lastly, all the numbers to the right of the rounded number are then replaced with zeros.

For example, a quiz question says to round 17,289,364 to the nearest ten thousand. After identifying the number in the ten thousand place, draw a tick mark over it.

$$\checkmark$$
$$17,289,364$$

Since the digit to the right of the 8 in the ten thousand place is "5 or more," the 8 is rounded up to 9.

$$9$$
$$17,289,364$$

The final step is to replace the numbers to the right of the rounded number with zeros.

$$0000$$
$$17,290,364$$

Therefore, rounding 17,289,364 to the nearest ten thousand results in 17,290,000.

Denominate Numbers

Despite its long name, denominate numbers are simple. A **denominate number** is a number with a unit of measurement attached. For example, 4 feet is a denominate number. The number 4, by itself, is not a denominate number. Once you add a description or unit to a number showing that it represents some kind of measurement, then it is classified as a denominate number.

The descriptive word *denominate* comes from the root word *denomination*. In this case, denomination means a group of units of measurement. Consider these other examples of denominate numbers: 6 gallons, 5³⁄₁₆″, 186 lb, 1800°F, and 55 miles per hour. Each number is attached to a unit of measurement, such as volume, length, mass, temperature, or speed.

As you may have guessed, welders work almost entirely with denominate numbers. This means that knowing how to work with denominate numbers is very important. There are certain rules to follow. For now, know the following. When an addition or subtraction question is expressed in denominate numbers, your answer must also be a denominate number.

Copyright by The Goodheart-Willcox Co., Inc.

Name _____ Date _____ Class _____

Unit 1 Practice

Label the place value of each digit in the following numbers.

1. 471

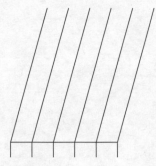

2. 15,286

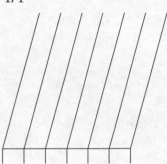

3. 349,015

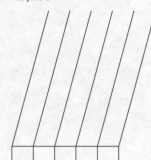

4. 27,636

Round the following numbers to the nearest hundred.

5. 694,701

6. 849

7. 17,213

8. 9,999

9. 29,897

10. 4,509

11. 945,999

12. 6,482,519

Round the following numbers to the nearest thousand:

13. 11,195

14. 449,561

15. 449,156

16. 85,028

17. 1,294,637

18. 9,195

19. 13,000,900

20. 10,001

Copyright by The Goodheart-Willcox Co., Inc.

21. In the list below, circle each denominate number.

    250 psi

    11,500 ft²

    five acres

    17.5 tons

    1,895

    0.525

    $7.65

    $7.65 per hour

    40″ per minute

    12

    12 dozen

Copyright by The Goodheart-Willcox Co., Inc.

Unit 2
Addition of Whole Numbers

Key Terms

addition   carrying

basic operations   sum

Introduction

Whole numbers are manipulated in arithmetic by four **basic operations**: addition, subtraction, multiplication, and division. The most widely used of these four operations is addition. You will learn about addition in this unit. The other three basic operations will be studied in following units.

Method Used to Add Whole Numbers

Addition is the process of combining two or more individual numbers to form a single number that usually has a higher value than any of the previously individual numbers. Whole numbers are added by stacking the numbers to be added in a column with all the units on the right side lined up one upon the other.

units	units	units
↓	↓	↓
4,956	27	354
21	192	2,032
191	11	111
256	1,542	9,461
+ 10,431	+ 5	+ 17

It is very, very important that all of the digits for each place value line up exactly above other digits of the same place value. Refer back to the illustration in the first unit on place values. Once each column has been calculated, a final answer is reached. The final answer in addition equations is called the **sum**.

Carrying

Only one digit can occupy a single column. What happens when the sum of numbers in a single column reaches or exceeds 10 and suddenly has a value that occupies two columns? In the example below, adding the numbers in the units column results in 15. The 5 is placed under the units column and the 1 is carried to the next column. Moving the second digit of a sum up a column in value is called **carrying**. Whenever a single column's sum totals a value of ten or more, move the higher valued numbers one column upward.

Copyright by The Goodheart-Willcox Co., Inc.

$$
\begin{array}{r}
^{2\,2\,1} \\
9{,}966 \\
21 \\
991 \\
256 \\
+\,94{,}531 \\
\hline
105{,}765
\end{array}
$$

When you carry a number, write the carried number at the top of the next column and make it smaller than your other digits. Next, add the tens column. The result is 26. Write in the number 6, and carry the 2 to the hundreds column. Continue in this manner until all columns are added. Always check your accuracy.

Look for Tens

When adding a column, search for combinations of numbers that add up to 10. In the single-column example below, 8 plus 2 equals 10, and 1 plus 9 equals 10 (those add up to 20, so far), and 4 plus 6 equals 10 (totaling 30). The remaining digit is 7, so the total is 37. You will find it easier doing the mental calculations this way. Be sure to tick off the numbers as you add them or you may forget which ones you have already added:

$$
\begin{array}{r}
9 \\
7 \\
2 \checkmark \\
6 \\
4 \\
1 \\
+\,8 \checkmark \\
\hline
37
\end{array}
$$

Checking Addition by Adding Up and Down

Always check your math work. If you added by going up the column, then check it by adding going down.

Mark the Answer

When you arrive at a final answer, mark it in a very clear manner. Try boxing it in, like this 248 or double underlining it 248. This is especially important if you have a mass of numbers on a sheet mixed in with previous, unrelated calculations. This is sometimes the situation when doing calculations in the shop or in the field.

Denominate Numbers

Just as important as marking the right number is including the right unit in the answer. When marking answers, be aware of denominate numbers. If a question is expressed in inches, your answer should also be in inches.

Copyright by The Goodheart-Willcox Co., Inc.

   When adding denominate numbers, be certain that all the numbers have the same unit. Only denominate numbers having the same type of unit can be added. If an addition equation contains denominate numbers with dissimilar units, decide which unit is the most convenient for addition. Convert the numbers to the convenient unit. Then add the numbers.

Compatible Units	**Incompatible Units**
4″	9 cm
+ 7″	+ 4″
11″	

The equation with 9 cm and 4″ cannot be calculated as it is. One of these denominate numbers must be converted to the other denominate number's unit of measurement. Then this equation can be completed. Conversion between different units will be covered later in the text.

   This concept of sharing common units is also applicable to subtraction equations. Denominate numbers must have the same units in order to subtract from one another. To refresh the concept of denominate numbers, review the Denominate Numbers section in Unit 1, *Introduction to Whole Numbers*.

Copyright by The Goodheart-Willcox Co., Inc.

Work Space

Copyright by The Goodheart-Willcox Co., Inc.

Name _____ Date _____ Class _____

Unit 2 Practice

Add the following number groupings. Show all your work. Be certain the columns line up. Box your answers.

1.	235	2.	6,241	3.	964
	471		7,356		33,508

4.	102	5.	75	6.	11,009
	268		5,491		7,489,621
	754		67,811		28,912,305
	888		604		47,951
	937		15,473		1,997

7.	8	8.	46	9.	167,540
	9		365		22,814
	7		3		499,863
	6		1,588		200
	2		331		98,157
	4		91		5
	1		942		37,254

Add the following number groupings. Box your answers.

10. $7 + 9 + 1 + 3 + 3 + 8 + 5 =$

11. $904 + 214 + 22 =$

12. $18,961 + 718 + 6,800 =$

13. $6,525 + 7,182 + 293 =$

14. $966 + 372 + 165 + 638 + 300 + 200 =$

15. $731 + 82 + 234 + 2,699 + 523 + 64 =$

Copyright by The Goodheart-Willcox Co., Inc.

16. 18 + 444 + 27,981 + 1,234 + 75,211 + 7 =

17. 21,987 + 101,001 + 622 + 9 + 36,299 + 981,202 =

18. 4 + 144 + 414,411 + 69,835 + 641,538 + 99 =

19. Three pieces of double extra-strong pipe in inventory have the following lengths: 27", 42", and 19". What is the total length of double extra-strong pipe in inventory?

20. Three welded frames are shipped by truck to a customer. The smallest frame weighs 985 lb, the next frame weighs 2,891 lb, and the largest weighs 3,257 lb. What is the total weight?

21. Tregaskiss, Inc., a manufacturer of welding rods, produced the following quantities: 9,622 for January, 72,450 for February, and 284,361 for March. What was the total production for the three months?

22. According to statistics, a coal mining company's production for each of the past six years has been 9,253,589 tons; 9,405,740 tons; 8,713,903 tons; 8,001,878 tons; 7,464,890 tons; and 7,791,261 tons. What was the total world production of coal for the past six years?

23. Alexander Lincoln worked the number of hours shown below for the month of December. What is the total number of hours he worked for the month?

December								
Week	Date	Hours	Week	Date	Hours	Week	Date	Hours
1	1	8	2	10	8	4	21	7
1	2	8	2	11	6	4	22	8
1	3	8				4	23	6
1	4	8	3	14	4	4	24	3
1	5	4	3	15	8			
			3	16	8	5	28	5
2	7	6	3	17	8	5	29	6
2	8	8	3	18	7	5	30	9
2	9	8	3	19	5	5	31	3

Copyright by The Goodheart-Willcox Co., Inc.

Name _____ Date _____ Class _____

Use the information below to answer the questions that follow. A manufacturing company with several departments is compiling information for their Human Resources records. The current list is as follows:

> 2,816 men and 1,042 women in the Fabricating Department
>
> 110 men and 119 women in the Repair and Maintenance Department
>
> 39 men and 63 women in the Administration Department
>
> 367 men and 457 women in the Engineering Department
>
> 98 men and 85 women in the Sales and Service Department

24. How many men are employed by the company?

25. How many women?

26. What is the total number of people employed?

27. What is the total length of angle iron in the illustration below?

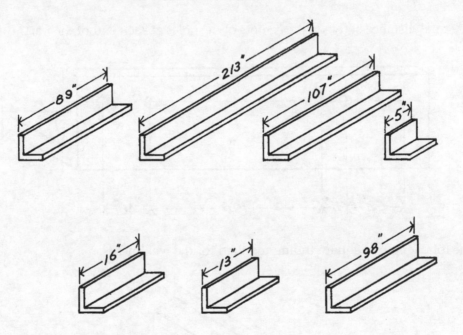

28. What is the total length of the notched plate below?

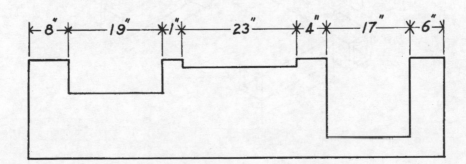

Copyright by The Goodheart-Willcox Co., Inc.

29. What is the total outside length of this angle iron frame below?

30. What is the total inside length of this frame?

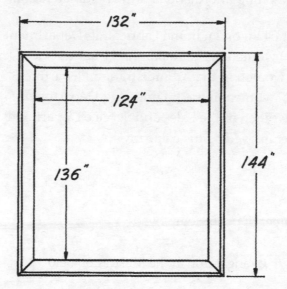

31. What is the total distance between the centers of the holes at each end of the part?

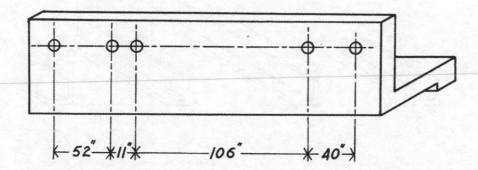

32. What is the total length of square tubing required for the weldment?

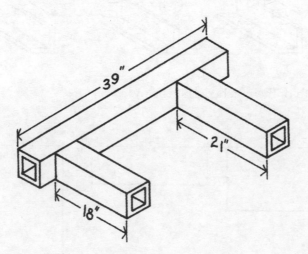

Copyright by The Goodheart-Willcox Co., Inc.

Name _____ Date _____ Class _____

33. This pie chart illustrates the number of hours required to complete various phases of Job Order #8805. What is the total number of hours required to complete the job?

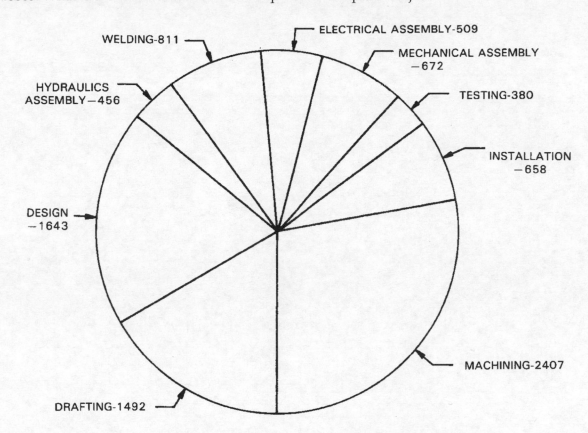

34. How many tons of steel are produced yearly by the countries listed below?

Annual Steel Production	
Country	**Production in Tons**
U.S.A.	67,746,000
Canada	14,825,000
China	35,620,000
Slovakia	15,319,000
France	21,346,000
Germany	41,662,000
Italy	25,085,000
Japan	101,708,000
Poland	15,112,000
Romania	13,193,000
Russia	148,980,000
U.K.	15,667,000

Copyright by The Goodheart-Willcox Co., Inc.

Work Space

Copyright by The Goodheart-Willcox Co., Inc.

Unit 3
Subtraction of Whole Numbers

Key Terms

borrow   minuend   subtraction
difference   remainder   subtrahend

Introduction

Subtraction is one of the four basic math operations. **Subtraction** is the process of finding the difference between two numbers. As with most math operations, there are new terms to learn. Below are the names of the parts of a subtraction equation:

$$1{,}347 \quad \leftarrow \textbf{Minuend}$$
$$\underline{-\ 26} \quad \leftarrow \textbf{Subtrahend}$$
$$1{,}321 \quad \leftarrow \textbf{Difference} \text{ or } \textbf{Remainder}$$

The top number in a subtraction equation is the **minuend**. Below it is the subtrahend. The **subtrahend** is the number that will be removed from the minuend. The resulting number is the **difference**, which may also be called the **remainder**.

Method Used to Subtract Whole Numbers

Two whole numbers are subtracted by stacking them one upon the other with the units columns lined up on the right side. The larger number belongs on the top position (minuend). Begin subtracting at the units column and work your way through the problem, column by column. Here are some examples:

$$\begin{array}{r} 84 \\ -\ 23 \\ \hline 1 \end{array} \qquad \begin{array}{r} 5{,}794 \\ -\ 62 \\ \hline 2 \end{array}$$

It is important to begin subtraction in the units column. Numbers in the lower columns can affect the values of the numbers in the higher columns. In our examples, the subtraction in the units columns has been completed successfully, allowing subtraction to continue in the higher columns.

$$\begin{array}{r} 84 \\ -\ 23 \\ \hline 61 \end{array} \qquad \begin{array}{r} 5{,}794 \\ -\ 62 \\ \hline 5{,}732 \end{array}$$

Copyright by The Goodheart-Willcox Co., Inc.

Borrowing

As previously mentioned, calculating subtraction equations must begin at the units column. This is because lower value columns can affect higher value columns. You will often encounter a situation like this:

$$\begin{array}{r} 47 \\ -\,19 \end{array}$$

The units column displays 7 minus 9. Since 9 cannot be subtracted from 7, you must **borrow** a number from the place value to the left. To borrow a number, subtract 1 from the number in the column to the left and add that 1 to the column that needs it. In this example, subtract 1 from the 4 in the tens column. Draw a line through the 4 and replace it with 3. Now, place a tiny 1 next to the 7 to create the number 17:

$$\begin{array}{r} {}^{3}\!\!\!\not{4}^{1}7 \\ -\,19 \end{array}$$

Now, subtract 9 from 17 to get 8. Then, to finish the problem, subtract 1 from 3 to get 2:

$$\begin{array}{r} {}^{3}\!\!\!\not{4}^{1}7 \\ -\,19 \\ \hline 28 \end{array}$$

Here is another more complicated example:

$$\begin{array}{r} 1{,}007 \\ -\,99 \end{array}$$

In the units column, you cannot subtract 9 from 7. As in the previous example, you look to the tens column to borrow a 1 and find there are no numbers to borrow. You then have to go to the hundreds column and still there is no number to borrow. Finally, in the thousands column, you can borrow a 1. By borrowing 1, the existing 1 becomes 0. Draw a line through the existing 1, and it becomes 0. Now, place a tiny 1 next to the 0 in the hundreds column to create the number 10.

$$\begin{array}{r} {}^{0}\!\!\not{1}^{1}{,}007 \\ -\,99 \end{array}$$

Now there is a 10 in the hundreds column. Borrow 1 from that 10 and reduce it to 9 by crossing out the 10 and replacing it with a 9. Add the borrowed 1 to the zero in the tens column.

$$\begin{array}{r} {}^{9}\!\!\not{0}\!\!\not{1}^{1}{,}007 \\ -\,99 \end{array}$$

Now, there is a 10 in the tens column. Borrow 1 from the 10 in the tens columns and reduce it to 9. Add 1 to the 7 in the units column, thereby changing it to 17. You can now begin the problem by subtracting 9 from 17 in the units column.

$$\begin{array}{r} {}^{9}\!\!\not{0}\!\!\not{1}\!\!\not{1}^{1}{,}007 \\ -\,99 \\ \hline 908 \end{array}$$

Copyright by The Goodheart-Willcox Co., Inc.

Checking Subtraction by Adding

Always check your accuracy. Check your subtraction work by adding the difference and the subtrahend. The total should equal the minuend:

Minuend	198,462
Subtrahend	− 76,291
Difference	122,171
Subtrahend	+ 76,291
Minuend	198,462

Denominate Numbers

Concerning denominate numbers, subtraction equations follow the same rule as addition equations. Denominate numbers in a subtraction equation must each have the same unit of measurement. If two denominate numbers with different units must be used in a subtraction equation, convert one of the units to the other.

Subtraction Terminology

When speaking or writing about subtraction, there are a number of ways of communicating the situation. If you are not familiar with these expressions, it could be difficult to figure out just what is being subtracted from what! Here are examples of how you are likely to see a subtraction equation stated:

24 take away 11

24 less 11

24 minus 11

24 subtract 11

24 reduced by 11

Find the difference between 24 and 11

All of these expressions are ways of writing the equation 24 − 11.

Copyright by The Goodheart-Willcox Co., Inc.

Work Space

Copyright by The Goodheart-Willcox Co., Inc.

Name _____   Date _____   Class _____

Unit 3 Practice

Subtract the following number groupings. Show all your work. Be certain the columns line up. Box your answers.

1.	875 − 211	2.	1,280 − 1,120	3.	9,706 − 7,960	
4.	74,240 − 659	5.	21,008 − 989	6.	12,345 − 6,789	
7.	45,000 − 2,370	8.	894,216 − 735,094	9.	1,019,471 − 4,048	

10. 847 − 305 =

11. 6,544 minus 822 =

12. Take 28,001 away from 29,477 =

13. Find the difference between 18,234 and 11,012 =

14. Find the result when 707 is taken away from 747 =

15. What is the difference between 74 and 47 =

16. Subtract 7,641 from 22,350 =

17. 184,555 less 90,915 =

18. Reduce 19,486 by 215 =

19. Advance Metal Works had 1,717 brackets in stock on February 15th. One month later, there were 912 remaining in inventory. How many brackets were used during the month?

Copyright by The Goodheart-Willcox Co., Inc.

Use this information to answer the questions that follow. A fabricating shop had the following steel in inventory:

      15,285 lb of 22 gage steel

      37,549 lb of 16 gage steel

      89,041 lb of 14 gage steel

One large customer order used the following quantities:

      6,102 lb of 14 gage steel

      2,819 lb of 16 gage steel

      11,910 lb of 22 gage steel

20. What weight of 22 gage steel remained after the order was completed?

21. What weight of 16 gage steel remained after the order was completed?

22. What weight of 14 gage steel remained after the order was completed?

23. What was the total weight of the customer order?

Use this information to answer the questions that follow. Proto Mfg. produced the following welding tips in one week:

Tip Size	Quantity Produced
#68	118,295
#51	7,050
#35	892

The following quantities were shipped to various welding supply houses:

Tip Size	Quantity Produced
#35	0
#68	73,460
#51	5,070

24. How many #68 tips remain at Proto?

25. How many #51 tips remain at Proto?

26. How many #35 tips remain at Proto?

Copyright by The Goodheart-Willcox Co., Inc.

Name _____ Date _____ Class _____

27. The original design for a large machine calls for a frame of structural steel tubing that weighs 4,715 lb. An engineering change was later approved, adding additional stiffeners. The final weight of the frame was 5,128 lb. What is the weight of the additional stiffeners?

28. A survey determined that there were 261,298 service stations in the U.S. After four years, a repeat of the survey showed this number had declined by 48,424. Among the stations still in business after four years, there were 54,631 that did not use welding equipment. How many service stations still used welding equipment at the time of the second survey?

29. A coal car was redesigned by replacing some of the steel with aluminum. The designers wanted to reduce its weight and also increase its payload to 110 tons. The total weight of the redesigned car was 42,695 lb. The amount of aluminum used was 9,860 lb. The remainder of the redesigned car was steel. What weight of steel was used?

30. Three pieces are cut from a length of cold-rolled steel. What length remains? Ignore loss of material due to cutting.

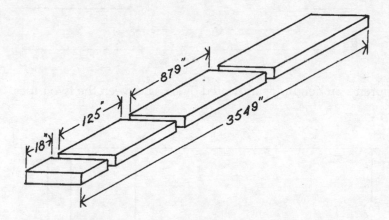

31. What is the distance between the centers of the holes in the piece below?

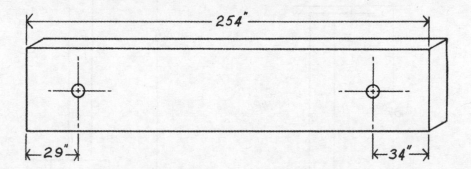

Copyright by The Goodheart-Willcox Co., Inc.

32. What is the height of the vertical piece (A)?

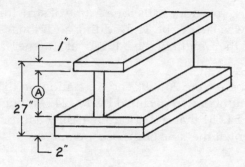

33. What is the length of distance (A)?

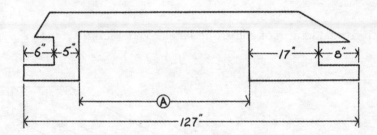

34. What is the difference in height (as indicated by (A)) between the two pipes?

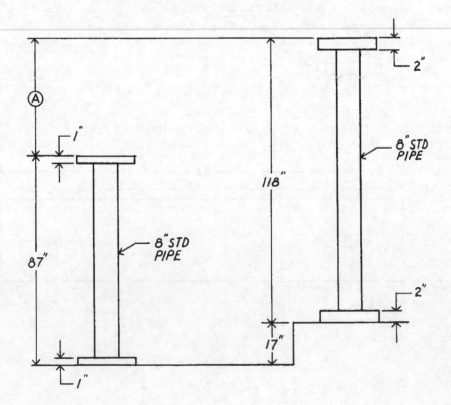

Copyright by The Goodheart-Willcox Co., Inc.

Name _____ Date _____ Class _____

Review the diagram and the supply room's pre-job inventory list.

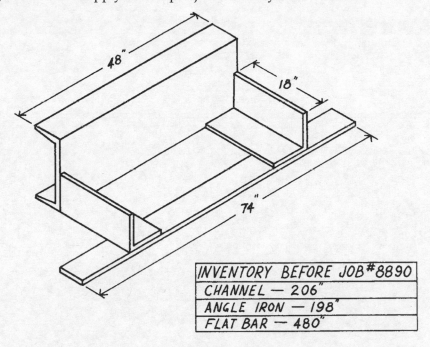

INVENTORY BEFORE JOB #8890
CHANNEL — 206"
ANGLE IRON — 198"
FLAT BAR — 480"

35. After this project is finished, how many inches of channel remain in the supply room?

36. After this project is finished, how many inches of angle iron remain in the supply room?

37. After this project is finished, how many inches of flat bar remain in the supply room?

Copyright by The Goodheart-Willcox Co., Inc.

Work Space

Copyright by The Goodheart-Willcox Co., Inc.

Unit 4
Multiplication of Whole Numbers

Key Terms

carry   multiplication   product
multiplicand   multiplier

Introduction

Multiplication can be thought of as a fast way to do addition. The process involves the repeated addition of a single number. This may seem odd at first, but look at it this way. If you were to add 3 + 3 + 3 + 3 + 3, the total, of course, would be 15. However, you could think of it as, "3 taken 5 times" or "5 times the number 3." This way of thinking works well for mental calculations of small numbers. It speeds up the process and simplifies the problem. Many people have memorized the multiplication facts for numbers 1 through 12 and these are often seen in chart form as multiplication tables.

Multiplication Table

2 × 1 = 2	3 × 1 = 3	4 × 1 = 4	5 × 1 = 5	6 × 1 = 6	7 × 1 = 7
2 × 2 = 4	3 × 2 = 6	4 × 2 = 8	5 × 2 = 10	6 × 2 = 12	7 × 2 = 14
2 × 3 = 6	3 × 3 = 9	4 × 3 = 12	5 × 3 = 15	6 × 3 = 18	7 × 3 = 21
2 × 4 = 8	3 × 4 = 12	4 × 4 = 16	5 × 4 = 20	6 × 4 = 24	7 × 4 = 28
2 × 5 = 10	3 × 5 = 15	4 × 5 = 20	5 × 5 = 25	6 × 5 = 30	7 × 5 = 35
2 × 6 = 12	3 × 6 = 18	4 × 6 = 24	5 × 6 = 30	6 × 6 = 36	7 × 6 = 42
2 × 7 = 14	3 × 7 = 21	4 × 7 = 28	5 × 7 = 35	6 × 7 = 42	7 × 7 = 49
2 × 8 = 16	3 × 8 = 24	4 × 8 = 32	5 × 8 = 40	6 × 8 = 48	7 × 8 = 56
2 × 9 = 18	3 × 9 = 27	4 × 9 = 36	5 × 9 = 45	6 × 9 = 54	7 × 9 = 63
2 × 10 = 20	3 × 10 = 30	4 × 10 = 40	5 × 10 = 50	6 × 10 = 60	7 × 10 = 70
2 × 11 = 22	3 × 11 = 33	4 × 11 = 44	5 × 11 = 55	6 × 11 = 66	7 × 11 = 77
2 × 12 = 24	3 × 12 = 36	4 × 12 = 48	5 × 12 = 60	6 × 12 = 72	7 × 12 = 84

(Continued)

Copyright by The Goodheart-Willcox Co., Inc.

8 × 1 = 8	9 × 1 = 9	10 × 1 = 10	11 × 1 = 11	12 × 1 = 12
8 × 2 = 16	9 × 2 = 18	10 × 2 = 20	11 × 2 = 22	12 × 2 = 24
8 × 3 = 24	9 × 3 = 27	10 × 3 = 30	11 × 3 = 33	12 × 3 = 36
8 × 4 = 32	9 × 4 = 36	10 × 4 = 40	11 × 4 = 44	12 × 4 = 48
8 × 5 = 40	9 × 5 = 45	10 × 5 = 50	11 × 5 = 55	12 × 5 = 60
8 × 6 = 48	9 × 6 = 54	10 × 6 = 60	11 × 6 = 66	12 × 6 = 72
8 × 7 = 56	9 × 7 = 63	10 × 7 = 70	11 × 7 = 77	12 × 7 = 84
8 × 8 = 64	9 × 8 = 72	10 × 8 = 80	11 × 8 = 88	12 × 8 = 96
8 × 9 = 72	9 × 9 = 81	10 × 9 = 90	11 × 9 = 99	12 × 9 = 108
8 × 10 = 80	9 × 10 = 90	10 × 10 = 100	11 × 10 = 110	12 × 10 = 120
8 × 11 = 88	9 × 11 = 99	10 × 11 = 110	11 × 11 = 121	12 × 11 = 132
8 × 12 = 96	9 × 12 = 108	10 × 12 = 120	11 × 12 = 132	12 × 12 = 144

1	2	3	4	5	6	7	8	9	10	11	12
2	4	6	8	10	12	14	16	18	20	22	24
3	6	9	12	15	18	21	24	27	30	33	36
4	8	12	16	20	24	28	32	36	40	44	48
5	10	15	20	25	30	35	40	45	50	55	60
6	12	18	24	30	36	42	48	54	60	66	72
7	14	21	28	35	42	49	56	63	70	77	84
8	16	24	32	40	48	56	64	72	80	88	96
9	18	27	36	45	54	63	72	81	90	99	108
10	20	30	40	50	60	70	80	90	100	110	120
11	22	33	44	55	66	77	88	99	110	121	132
12	24	36	48	60	72	84	96	108	120	132	144

To use this chart, the product of multiplying each number along the top row by each number in the left column is found where the row and column intersect. For example, to find the answer to 8 times 4, locate their intersecting squares. Following column 8 down to row 4 shows the answer to be 32. Likewise, following column 4 down to row 8 also results in 32.

Incidentally, there is no special reason why the multiplication table stopped at 12. It is probably because most people did not want to memorize beyond that point. Even in this age of electronic calculators, you should be very familiar with the multiplication table. You should commit it to memory.

Method Used to Multiply Whole Numbers

Large numbers are calculated by writing them out. To begin, you should know the appropriate terminology.

$$
\begin{array}{r}
132 \\
\times\,23 \\
\hline
3{,}036
\end{array}
$$

    132 ← **Multiplicand**

  × 23 ← **Multiplier**

 3,036 ← **Product**

Copyright by The Goodheart-Willcox Co., Inc.

The upper number in a multiplication equation is the **multiplicand**. This is the number that is being added repeatedly. The number below the multiplicand is the **multiplier**. This number represents how many times the multiplicand will be repeatedly added. The result of a multiplication equation is the **product**. Refer to the multiplication tables if you need to refresh your skills as you study the following examples.

Always line up the figures on the right side in the units column. Multiply every digit in the multiplicand by the number in the units column of the multiplier. Start multiplying the multiplicand at the units column. In this example, it starts with the 3 in the multiplier's units column and the 2 in the multiplicand's units column.

$$
\begin{array}{r}
132 \\
\times\,23 \\
\hline
396
\end{array}
$$

Next, multiply every digit in the multiplicand by the 2 in the multiplier's tens column. Place the results of this multiplication below the first results. However, these second results should be indented so it lines up directly under the 2 in the multiplier. Then, add the results of the two multiplication operations.

$$
\begin{array}{r}
132 \\
\times\,23 \\
\hline
396 \\
264 \\
\hline
3{,}036
\end{array}
$$

In multiplication problems, you often have to **carry** numbers. This was also done in addition equations when the sum of a single column was 10 or higher. Since only one digit can occupy a column, a number was carried to the next column. See the example below.

$$
\begin{array}{r}
298 \\
\times\,27
\end{array}
$$

In the first operation, $7 \times 8 = 56$. Place the 6 in the answer and write a little 5 directly above the 9.

$$
\begin{array}{r}
\overset{5}{} \\
298 \\
\times\,27 \\
\hline
6
\end{array}
$$

Now, multiply 7×9 to get 63 and add the carried 5 to it, arriving at 68. Place 8 in the answer and a little 6 above the 2.

$$
\begin{array}{r}
\overset{65}{} \\
298 \\
\times\,27 \\
\hline
86
\end{array}
$$

Multiply 7×2 to get 14. Add the carried 6 to the 14 to get 20. Write 20 in the answer. Always check your accuracy.

$$
\begin{array}{r}
\overset{65}{} \\
298 \\
\times\,27 \\
\hline
2086
\end{array}
$$

Copyright by The Goodheart-Willcox Co., Inc.

For larger equations like this one, neatness is important. When multiplying the next set of numbers, write the carried numbers above the previously used carried numbers. Keep careful track of these numbers, so you do not add the wrong numbers. In larger equations, you may want to cross out carried numbers after they have been used in a calculation.

$$
\begin{array}{r}
11 \\
\cancel{65} \\
298 \\
\times\ 27 \\
\hline
2086 \\
596 \\
\hline
8{,}046
\end{array}
$$

Accurate Alignment

Be sure to line up the numbers accurately in your answer. You may include right-hand zeros or X's to help keep the figures aligned. These zeros and X's occupy the space of indented columns.

$$
\begin{array}{r}
154 \\
\times\ 368 \\
\hline
1232 \\
9240 \\
46200 \\
\hline
56{,}672
\end{array}
\qquad
\text{Zeros added to keep columns aligned.}
\qquad\qquad
\begin{array}{r}
154 \\
\times\ 368 \\
\hline
1232 \\
924\text{X} \\
462\text{XX} \\
\hline
56{,}672
\end{array}
\qquad
\text{X's added to keep columns aligned.}
$$

Notice the emboldened zeros and Xs in the examples above. Including such characters will help to keep your columns properly aligned. Leaving those spaces open may lead to column confusion and incorrect results.

Checking Multiplication by Reversing Positions

A method of checking your multiplication work is to reverse the positions of the multiplier and multiplicand and repeat the problem. Performing the equation each way should result in the same product. The numbers obtained by multiplying each digit of the multiplier with the multiplicand will be different than the numbers from the first equations. However, these different numbers should still add up to the same final product as the original equation.

Original Equation		Reverse Position Equation
643		972
× 972		× 643
1286	←— Different —→	2916
45010	←— Different —→	38880
578700	←— Different —→	583200
624,996	←— Same —→	624,996

As shown in the example above, the initial multiplication results may be different, but the final product should match the original answer.

Copyright by The Goodheart-Willcox Co., Inc.

Smaller Multiplier

It does not matter which of the two numbers is positioned as the multiplier or multiplicand. Both will provide the same results. However, it may be easier to place the smaller number as the multiplier. This will result in having fewer rows of numbers to add.

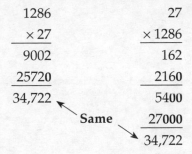

Denominate Numbers

How denominate numbers work in an equation is based on the operation of an equation. For instance, we already know that denominate numbers must have the same units to be added or subtracted. However, in multiplication and division equations, denominate numbers do not need to have the same units. In a multiplication equation, the units of measurements are combined.

When denominate numbers with different units of measurement are multiplied together, they form a new unit of measurement. See the example below.

$$12 \text{ lb} \times 7' = 84 \text{ ft-lb}$$

While the numbers are multiplied, the pound and foot units are combined to form foot-pounds (ft-lb). These compound units can be formed by multiplication and undone by division. When denominate numbers sharing the same unit of measurement are multiplied together, they form a squared unit. See the example below.

$$7'' \times 8'' = 56 \text{ in}^2$$

Multiplying two inch measurements together forms a product measured in inches squared (in^2). More information on denominate numbers in multiplication equations will be covered in later units on measurement.

Copyright by The Goodheart-Willcox Co., Inc.

Work Space

Copyright by The Goodheart-Willcox Co., Inc.

Name _____   Date _____   Class _____

Unit 4 Practice

Multiply the following number groupings. Show all your work. Be certain the columns line up. Box your answers.

1.   12
        $\times\,34$

2.   98
        $\times\,75$

3.   56
        $\times\,70$

4.   659
       $\times\,423$

5.   207
       $\times\,845$

6.   913
       $\times\,600$

7.   84,796
       $\times\,56$

8.   93,253
       $\times\,24$

9.   53,602
       $\times\,35$

10. $253 \times 6{,}944 =$

11. $804 \times 5{,}022 =$

12. $68 \times 12{,}673 =$

13. $69{,}671 \times 36 =$

14. $12 \times 24 \times 36 =$

15. $18 \times 13 \times 20 =$

16. $109 \times 14 \times 34 =$

17. $263 \times 101 \times 24 =$

18. $468 \times 219 \times 153 =$

19. Lee's Welding produced 193 base frames. Each frame required 98 spot welds. What is the total number of spot welds for the entire job?

20. Ramido's Rapid Delivery Service delivered 4 pallets of rivets to Weld-Can Mfg. Each pallet contained 27 cartons and each carton contained 24 boxes. Each box contained 68 rivets. How many rivets were delivered?

Copyright by The Goodheart-Willcox Co., Inc.

21. Thirty-nine rows of studs will be welded to a metal platform. Each row will be 5″ apart. There will be 27 studs in each row. What is the total number of studs required?

22. A GMAW welder traveling at 17″ per minute was in constant use for 17½ hours each day for 24 days. How many inches of weld were deposited in that time?

23. The roof of an arena built for the University of Iowa requires 5,247 plates to secure the roof trusses to the support beams. These plates vary in size and number of bolts required. The plates were flame cut to shape, then subcontracted to a machine shop for drilling the bolt holes. According to the information listed below, what is the total number of bolt holes that need to be drilled?

    1,728 plates with 6 holes   2,235 plates with 14 holes

    295 plates with 5 holes   15 plates with 23 holes

    964 plates with 9 holes   10 plates with 10 holes

Use the information below to answer the questions that follow. The supplies listed below must be transported from the ground floor storage to the top floor. Workers plan to use a freight elevator with a capacity of 8,500 lb to send everything at once.

Steel Rods

Seven bundles of 17 rods each. Each rod is 4′ long and weighs 2 lb per foot.

Plates

Twelve stacks of 13 plates each. Each plate weighs 14 lb. The plates are stacked on 4 wooden pallets that weigh 39 lb each.

Angle Iron

Forty-two pieces. Each piece weighs 18 lb and is 4′ long.

Aluminum Tubing

Six bundles of 16 tubes. Each tube weighs 7 lb.

Four bundles of 12 tubes. Each tube weighs 9 lb.

Seven bundles of 5 tubes each. Each tube is 4′ long and weighs 2 lb per foot.

Four bundles of 5 tubes each. Each tube is 3′ long and weighs 2 lb per foot.

Pipe

Two bundles of 20 pipes each. Each pipe is 3′ long and weighs 2 lb per foot.

Seven bundles of 11 pipes each. Each pipe weighs 11 lb.

Hubs

Four cartons of 19 boxes each. Each box contains 6 hubs. The weight of each box including the 6 hubs is 13 lb. Each carton is on a wooden pallet that weighs 39 lb.

Operator

The elevator operator weighs 165 lb.

24. Will the workers be able to send the whole load in a single trip?

25. What is the total load?

Copyright by The Goodheart-Willcox Co., Inc.

Name _____ Date _____ Class _____

Use the information below to answer the questions that follow. A power line built for Louisiana Light and Power Co. had 619 towers constructed of various sizes of pipe. Specifications for the job are listed below.

Steel Pipe Requirements			
	Tower Legs	Cross Arms	Stiffeners
Material	ASTM A595	ASTM A595	ASTM A595
Size	2′ diameter	1½′ diameter	1′ diameter
Number per Tower	2	1	2
Length	93′ on 175 towers	102′	35′
	108′ on 300 towers		
	122′ on 144 towers		

26. Calculate the total length of 2′ diameter pipe.

27. Calculate the total length of 1½′ diameter pipe.

28. Calculate the total length of 1′ diameter pipe.

29. Find the total length of pipe for the entire project.

Use the information below to answer the questions that follow. The illustration below depicts equally-sized rings that are welded together.

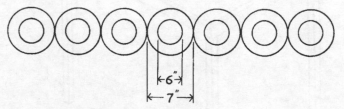

30. What is the total length of the weldment?

31. What is the total weight of the weldment if each ring weighs 6 lb?

Copyright by The Goodheart-Willcox Co., Inc.

32. How many of the following GMAW welding tips can be produced in 67 working days at 9 hours per day?

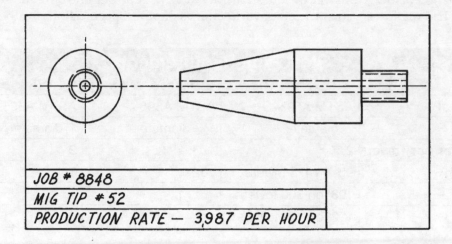

JOB # 8848

MIG TIP #52

PRODUCTION RATE — 3,987 PER HOUR

33. How many inches of chrome plated tubing will be required to manufacture 289 tables?

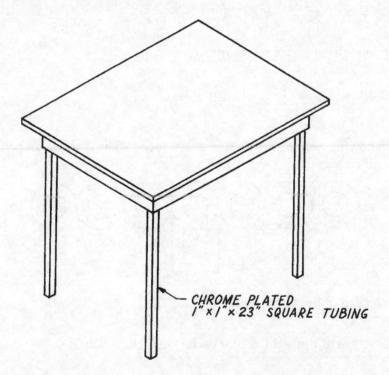

CHROME PLATED
1" x 1" x 23" SQUARE TUBING

Copyright by The Goodheart-Willcox Co., Inc.

Name _____ Date _____ Class _____

34. Each of the posts in this shaft support is welded all around. How many inches of weld will be applied to make 12,983 shaft supports?

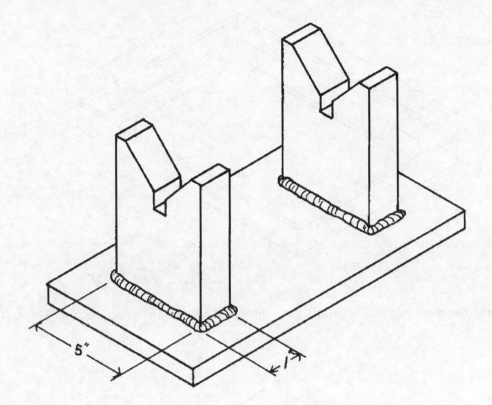

35. Review the diagram below. Calculate the weight of the weldment. Assume there is no filler added to the welds.

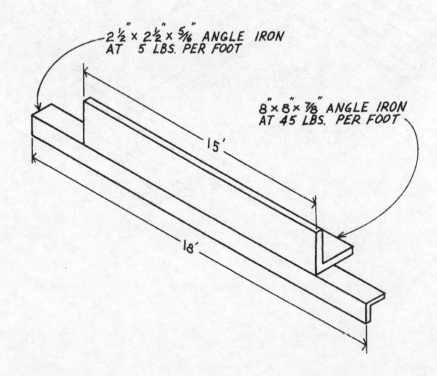

Copyright by The Goodheart-Willcox Co., Inc.

36. Review the diagram below. Calculate the weight of the weldment. Assume there is no filler added
     to the welds.

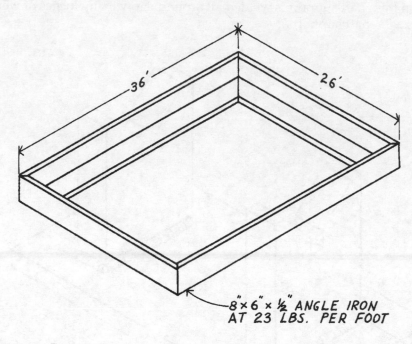

8"× 6"× ½" ANGLE IRON
AT 23 LBS. PER FOOT

37. Review the diagram below. Calculate the weight of the weldment. Assume there is no filler added
     to the welds.

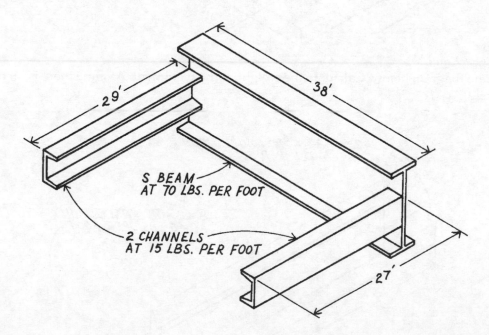

S BEAM
AT 70 LBS. PER FOOT

2 CHANNELS
AT 15 LBS. PER FOOT

Copyright by The Goodheart-Willcox Co., Inc.

Name _____ Date _____ Class _____

38. The diagram below shows the measurements for a frame. Eighty-five frames are to be fabricated. What is the total number of inches of tubing required?

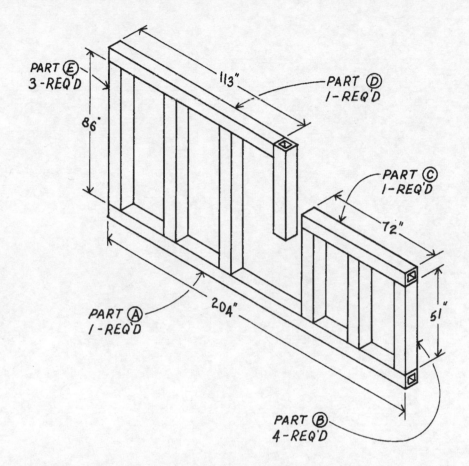

PART E
3-REQ'D

113"

PART D
1-REQ'D

86"

PART C
1-REQ'D

72"

PART A
1-REQ'D

204"

51"

PART B
4-REQ'D

Copyright by The Goodheart-Willcox Co., Inc.

Work Space

Copyright by The Goodheart-Willcox Co., Inc.

Unit 5
Division of Whole Numbers

Key Terms

dividend   divisor   remainder

division   quotient

Introduction

Division is the process of finding how many times one number can be "contained" in another number. For example, the number 50 "contains" the number 10, five times. There are a variety of expressions used to describe this relationship.

There are 5 tens in 50.

Ten "goes into" 50, five times.

Fifty divided by 10 equals five.

Ten divides into 50, five times.

While there are a number of ways of expressing this in spoken English, there are also a number of ways of symbolizing division. All of the following symbolize the dividing of 1,188 by 9.

$$9\overline{)1,188}$$

$$1,188 \div 9$$

$$\frac{1,188}{9}$$

$$1,188/9$$

Method Used to Divide Whole Numbers

To begin, you should know the appropriate terminology. These words (like many math terms) may seem awkward at first. You will see them mainly in math books and occasionally in some formula tables. They are needed to help provide a smooth explanation of the division process.

$$132 \leftarrow \textbf{Quotient}$$
$$\textbf{Divisor} \rightarrow 9\overline{)1,188} \leftarrow \textbf{Dividend}$$

The **dividend** is the number that will be divided into smaller equal units. It is the largest number in the equation. The **divisor** is the number of times the dividend will be broken into equal pieces. The **quotient** is the number of the equally sized units into which the dividend was broken by the divisor.

1. To begin a division operation, compare the first digit of the dividend with the divisor. Cover all of the dividend, except the leftmost digit, with your finger.

$$7\overline{)131,138}$$

Copyright by The Goodheart-Willcox Co., Inc.

2. Then, ask yourself, "Will the divisor go into this digit?" If yes, then ask how many times. If the answer is "No", as in this example, uncover the next digit and ask if the divisor will divide into those digits of the dividend and how often. In this example, the answer is "Yes." The divisor 7 will divide into 13, once. Write the 1 above the 3 as shown. It is extremely important to line up the numbers accurately.

$$\begin{array}{r} 1 \\ 7\overline{)131,138} \end{array}$$

3. After writing the 1 above the 3, multiply 7×1 and write the answer under the 13.

$$\begin{array}{r} 1 \\ 7\overline{)131,138} \\ 7 \end{array}$$

4. Next step, subtract 7 from 13.

$$\begin{array}{r} 1 \\ 7\overline{)131,138} \\ \underline{-7} \\ 6 \end{array}$$

5. "Borrow" or "bring down" the next digit in the dividend. Draw an arrow from that digit in the dividend pointing to the space to the right of the remaining number. Write that digit beside the remaining number. The 6 becomes a 61. This arrow from the dividend to the remaining number will show from where each digit is borrowed in the dividend. These arrows are especially useful in long division and will keep columns neatly aligned.

$$\begin{array}{r} 1 \\ 7\overline{)131,138} \\ \underline{-7\downarrow} \\ 61 \end{array}$$

6. Again, compare the divisor and leftmost dividend digits. "Does the divisor go into 61?" The answer is "Yes," eight times. Write an 8 directly above the leftmost 1 in the dividend. Multiply 7×8 to get 56 and write this product beneath the 61. Perform subtraction.

$$\begin{array}{r} 18 \\ 7\overline{)131,138} \\ \underline{-7} \\ 61 \\ \underline{-56} \\ 5 \end{array}$$

7. Continue repeating this cycle with each of the remaining digits.

$$\begin{array}{r} 18,734 \\ 7\overline{)131,138} \\ \underline{-7} \\ 61 \\ \underline{-56\downarrow} \\ 5\,1 \\ \underline{-49\downarrow} \\ 23 \\ \underline{-21\downarrow} \\ 28 \\ \underline{-28} \\ 0 \end{array}$$

Copyright by The Goodheart-Willcox Co., Inc.

Division Equation Variations

The previous example had a dividend and divisor that divided easily. Not all problems proceed so smoothly. There are two variations you will encounter. Firstly, there will be some occasions where the divisor cannot divide into the number that has been "brought down."

$$
\begin{array}{r}
1 \\
7\overline{)728} \\
-7\downarrow \\
\hline
02
\end{array}
$$

The number 2 has already been "brought down," but will the divisor go into 2? The answer is "No." In this case, place a zero above the 2 in the dividend and bring down the next number.

$$
\begin{array}{r}
10 \\
7\overline{)728} \\
-7\downarrow \\
\hline
028
\end{array}
$$

Forgetting to place this zero in the quotient is an error that a beginner can easily make. Now, proceed with the problem.

$$
\begin{array}{r}
104 \\
7\overline{)728} \\
-7 \\
\hline
028 \\
-28 \\
\hline
0
\end{array}
$$

The second division equation variation is when the divisor does not divide evenly into the dividend. In these cases, you will have a **remainder**. This is a number remaining after final subtraction of divisor and dividend digit products.

$$
\begin{array}{r}
259\ \text{r}\ 1 \\
5\overline{)1,296} \\
-10\downarrow \\
\hline
29 \\
-25\downarrow \\
\hline
46 \\
-45 \\
\hline
1
\end{array}
$$

The remainder here is 1. Include the remainder with the rest of the quotient but separate it with an "r." Always check your accuracy.

Checking Division by Multiplying

Check your work by multiplying the quotient by the divisor. If there was a remainder, add it to the product and the answer should be the same as the dividend. The equation below checks the accuracy of remainder problem example.

$$
\begin{array}{r}
259 \\
\times 5 \\
\hline
1295 \\
+1 \\
\hline
1,296
\end{array}
$$

Copyright by The Goodheart-Willcox Co., Inc.

Do Not Crowd

As you can see, division problems tend to use a lot of space. Always give yourself plenty of room. Do not try to squeeze the figures into a small space. Trying to save space will only increase your chances of making an error.

Denominate Numbers

As in multiplication equations, denominate numbers in division equations do not need to have the same unit. However, the divisor's unit is one part of the combined unit of the dividend. In a division equation of denominate numbers, the divisor's unit of measurement can cancel out the part of the dividend's unit of measurement that they share in common. In this way, compound units consisting of two different units can be separated into their distinct parts.

$$
\begin{array}{r}
25' \\
5\ \text{lb} \overline{)\ 125\ \text{ft-lb}} \\
-10 \\
\hline
25 \\
-25 \\
\hline
0
\end{array}
$$

In dividing 125 ft-lb by 5 lb, treat the numbers like a normal division equation. However, the units are handled differently. The pound (lb) unit cancels out pound (lb) in the foot-pound (ft-lb) unit, resulting in feet (ft).

Squared units can be viewed as a type of compound unit. In calculating a division equation, a squared unit can be divided into a single unit.

$$
\begin{array}{r}
6'' \\
12\ \text{in} \overline{)\ 72\ \text{in}^2} \\
-72 \\
\hline
0
\end{array}
$$

In dividing 72 in^2 by 12″, treat the numbers like a normal division equation. The inch unit in the divisor cancels one of the inch units in the dividend (in^2 is the same as in-in), resulting in inch (in) unit for the quotient. These concepts will be reinforced in later units on measurement.

Copyright by The Goodheart-Willcox Co., Inc.

Name _____ Date _____ Class _____

Unit 5 Practice

Divide the following number groupings. Show all your work. Be certain the columns line up. Box your answers.

1. $8\overline{)216}$ 2. $29\overline{)348}$ 3. $6\overline{)618}$

4. $7\overline{)8,050}$ 5. $9\overline{)81,064}$ 6. $35\overline{)2,410}$

7. $143\overline{)1,001}$ 8. $365\overline{)44,895}$ 9. $400\overline{)160,000}$

10. $4,826 \div 3,519 =$

11. 1,840 divided by 72 =

12. $1,066\overline{)981,734}$

13. $18,604/290 =$

14. $1,776\overline{)3,528,912}$

15. 245 divided into 6,195 =

16. $8,521\overline{)340,915,264}$

17. How many 335's are there in 6,369?

18. Divide 648 into 8,759

19. A company that manufactures electric light fixtures used 16,653 lb of solder in one year. If there were 273 working days in the year, how many pounds of solder would be consumed on an average day?

20. Welding a heavy steel plate with a V-groove joint required 1,248 ounces of filler. The plate was 3" thick and 208" long. How many ounces of filler were used for each inch of weld?

21. A piece of angle iron 120" long is to be cut into 19" pieces. Assuming there is no loss of material due to cutting, how many 19" long pieces would be obtained?

Copyright by The Goodheart-Willcox Co., Inc.

Use this information to answer the questions that follow. A fabricating shop was successful in their bid for the contract to supply railings for a bridge across the Detroit River. The bridge was to be 4,784 long and railings were to be provided for both sides of the bridge. The railings were welded in 16' sections and delivered to the job site.

22. How many sections of railing were built?

23. If each section weighed 642 lb, what was the total weight of the railings?

24. A supply house ordered 14 cartons of aluminum flux in cans. Each fully packed carton weighed 216 lb (excluding the weight of the carton). If each can weighed 3 lb, how many cans of flux are in the order?

25. During a recent repair of the Golden Gate Bridge, 30,960' of on-site welding was laid. The job was completed in 169 days. What was the average number of feet of weld laid per day? Express your answer to the nearest foot.

26. A job calls for clips to be welded at each end of a 1,863' seawall. Clips are also to be welded 18" apart along the length of the wall. How many clips will be used on this job?

27. A 96" piece of flat bar is to have 11 center lines scribed at equal distances apart. How far apart are the center lines? Hint: Draw a sketch of the bar and the center lines.

28. What is the distance between the centers of the evenly spaced holes shown below?

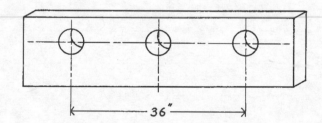

29. What is the distance between the centers of the evenly spaced holes?

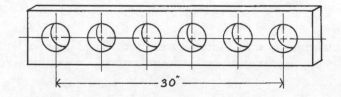

Copyright by The Goodheart-Willcox Co., Inc.

Name _____ Date _____ Class _____

30. The job order for the weldment below calls for it to have 37 holes drilled evenly spaced apart. Calculate the distance between the holes.

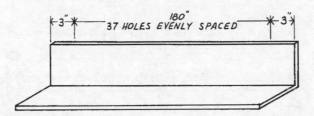

31. How many 17″ pieces can be cut from this 16,761″ roll of steel?

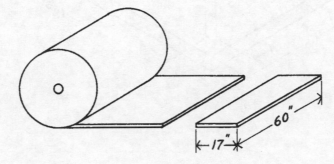

32. Illustrated is one of eight bundles of round bar stock that make up a delivery weighing 14,208 lb. What is the weight of each bar?

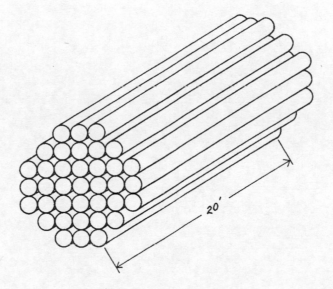

Copyright by The Goodheart-Willcox Co., Inc.

33. Studs are to be welded to this plate in the following manner. Studs along the length are to be
    spaced 9″ apart and studs along the width are to be spaced 8″ apart. How many studs are required
    to fill the plate with studs?

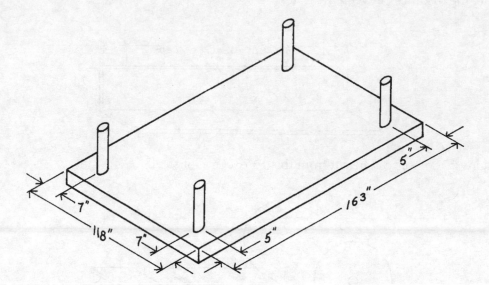

Section 2

Common Fractions

Section Objectives

After studying this section, you will be able to:

- Explain the parts of a fraction
- Provide examples of proper fractions, improper fractions, and mixed numbers
- Demonstrate how to reduce equivalent fractions
- Demonstrate how to convert between mixed numbers and improper fractions
- Perform addition of fractions
- Perform subtraction of fractions
- Perform multiplication of fractions
- Perform division of fractions

Copyright by The Goodheart-Willcox Co., Inc.

Unit 6
Introduction to Common Fractions

Key Terms

common fraction	higher terms	lowest terms	proper fraction
denominator	improper fraction	mixed number	reducing
equivalent fraction	lower terms	numerator	terms

Introduction

When you divide something, such as a steel rod, into a number of equal pieces, each of those pieces can be called a fraction of the whole rod. If the rod is cut into ten equal parts, each part is one-tenth of the whole. This can be expressed in numbers by the fraction $\frac{1}{10}$. A fraction such as this is properly called a **common fraction**. In everyday usage, it is often referred to as a fraction. Three of the parts make up $\frac{3}{10}$ of the whole, 9 parts would be $\frac{9}{10}$, and finally, all ten parts could be written as $\frac{10}{10}$. Since all ten parts make up the whole, you can see that $\frac{10}{10} = 1$. Any fractions containing all the parts are equal to 1: $\frac{64}{64}$, $\frac{17}{17}$, $\frac{256}{256}$, $\frac{1}{1}$, $\frac{12}{12}$.

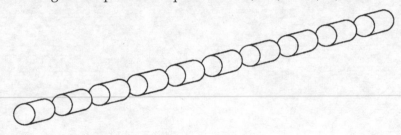

The following three illustrations show fractional parts of an inch.

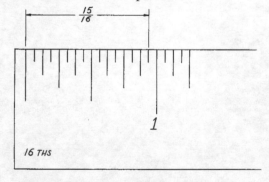

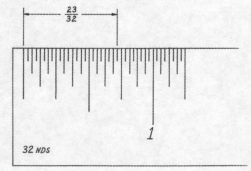

Copyright by The Goodheart-Willcox Co., Inc.

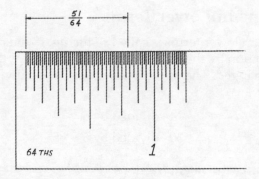

Parts of a Fraction

The parts of a fraction are shown below.

$$\dfrac{3}{4} \begin{array}{l} \leftarrow \textbf{numerator} \\ \leftarrow \textbf{denominator} \end{array}$$

An easy way to remember these terms and to keep them separate is to think that the <u>d</u>enominator is <u>d</u>own on the bottom. Associate the "d" in <u>d</u>enominator with the "d" in <u>d</u>own. The numerator and denominator together are referred to as the **terms** of the fraction. The **denominator** refers to how many equal parts the item has been divided into. The **numerator** refers to how many of those parts you are using.

Proper Fractions

A **proper fraction** has a numerator smaller than the denominator. Examples include the following: $\frac{1}{4}$, $\frac{18}{57}$, $\frac{9}{32}$.

Improper Fractions

An **improper fraction** has a numerator larger than the denominator. These fractions are "top heavy." Examples include the following: $\frac{15}{7}$, $\frac{29}{17}$, $\frac{9}{8}$.

Mixed Numbers

A **mixed number** consists of a whole number and a fraction. Examples include the following: $1\frac{1}{2}$, $37\frac{11}{16}$, $108\frac{5}{13}$.

Equivalent Fractions

Equivalent fractions are fractions that are equal in value to each other. Any fraction can be expressed as an equivalent fraction in **higher terms** or **lower terms**.

Equivalent Fractions in Higher Terms

To express a fraction in higher terms, multiply the numerator and denominator of a fraction by the same number. Since both terms were raised by the same number, the value of the two fractions will be the same, as illustrated:

$$\dfrac{3 \text{ multiplied by } 7 =}{4 \text{ multiplied by } 7 =} \dfrac{21}{28}$$

The fractions $\frac{3}{4}$ and $\frac{21}{28}$ are equivalent fractions.

Copyright by The Goodheart-Willcox Co., Inc.

Equivalent Fractions in Lower Terms

Expressing a fraction in lower terms is called **reducing**. If you divide both the numerator and denominator of a fraction by the same number, the fraction will be reduced. The value of the fraction will not be changed. The numbers, of course, must divide evenly (that is, without a remainder) as illustrated:

$$\frac{24 \text{ divided by } 6 =}{30 \text{ divided by } 6 =} \frac{4}{5}$$

The fractions ²⁴⁄₃₀ and ⁴⁄₅ are equivalent fractions.

Here is another example:

$$\frac{42}{56} = \frac{6}{8} = \frac{3}{4}$$

When a fraction cannot be reduced further, it is considered to be in its **lowest terms**. When you arrive at the final answer, you are expected to reduce the answer to its lowest terms.

Reducing Improper Fractions

Improper fractions are reduced by dividing the numerator by the denominator. If there is a remainder, it becomes the new numerator of the fraction. See the examples below.

$$\frac{19}{4} \qquad \frac{22}{10} \qquad \frac{35}{5}$$

$$4\overline{)19} \qquad 10\overline{)22} \qquad 5\overline{)35}$$

$$\begin{array}{r} 4\text{ r }3 \\ 4\overline{)19} \\ -16 \\ \hline 3 \end{array} \qquad \begin{array}{r} 2\text{ r }2 \\ 10\overline{)22} \\ -20 \\ \hline 2 \end{array} \qquad \begin{array}{r} 7 \\ 5\overline{)35} \\ -35 \\ \hline 0 \end{array}$$

$$4\frac{3}{4} \qquad\qquad 2\frac{1}{5} \qquad\qquad 7$$

Changing Mixed Numbers to Improper Fractions

This is an operation useful in solving certain math problems you will encounter. For this example, we will use the mixed number: 7⅔.

1. Firstly, multiply the whole number by the denominator.

$$7 \times 3 = 21$$

2. Next, add the numerator to the value of the whole number and denominator product.

$$21 + 2 = 23$$

3. Lastly, place this sum over the denominator of the original mixed number.

$$7\frac{2}{3} = \frac{23}{3}$$

Copyright by The Goodheart-Willcox Co., Inc.

Name _____ Date _____ Class _____

Unit 6 Practice

Express the following as equivalent fractions. Show all your work. Box your answers.

1. $\dfrac{5}{8} = \dfrac{?}{16}$

2. $\dfrac{1}{2} = \dfrac{?}{200}$

3. $\dfrac{1}{1} = \dfrac{?}{98}$

4. $\dfrac{20}{64} = \dfrac{?}{16}$

5. $\dfrac{28}{32} = \dfrac{?}{8}$

6. $\dfrac{128}{256} = \dfrac{?}{2}$

7. $\dfrac{11}{13} = \dfrac{?}{65}$

8. $\dfrac{3}{19} = \dfrac{?}{133}$

9. $\dfrac{4}{5} = \dfrac{?}{100}$

Determine whether the following pairs of fractions are equivalent fractions. Answer each group with *yes* or *no*.

10. $\dfrac{2}{3}$ and $\dfrac{65}{99}$

11. $\dfrac{3}{4}$ and $\dfrac{21}{28}$

12. $\dfrac{17}{64}$ and $\dfrac{17}{128}$

13. $\dfrac{1}{1}$ and $\dfrac{2}{1}$

14. $\dfrac{10}{11}$ and $\dfrac{110}{121}$

15. $\dfrac{19}{33}$ and $\dfrac{76}{132}$

16. $\dfrac{11}{111}$ and $\dfrac{99}{999}$

17. $\dfrac{33}{19}$ and $\dfrac{132}{76}$

18. $\dfrac{100}{10}$ and $\dfrac{10}{100}$

Reduce the following improper fractions. Show all your work. Box your answers.

19. $\dfrac{18}{4}$

20. $\dfrac{111}{11}$

21. $\dfrac{141}{8}$

22. $\dfrac{1968}{16}$

23. $\dfrac{7}{2}$

24. $\dfrac{75}{25}$

25. $\dfrac{1505}{150}$

26. $\dfrac{10830}{833}$

27. $\dfrac{2}{1}$

Copyright by The Goodheart-Willcox Co., Inc.

Change the following mixed numbers to improper fractions. Show all your work. Box your answers.

28. $1\dfrac{1}{2}$ 29. $12\dfrac{3}{4}$

30. $17\dfrac{11}{64}$ 31. $29\dfrac{1}{8}$

32. $37\dfrac{3}{7}$ 33. $185\dfrac{5}{9}$

34. $13\dfrac{11}{13}$ 35. $10\dfrac{10}{10}$

36. $7\dfrac{7}{9}$

Express the following fractions in the lowest terms. Show all your work. Box your answers.

37. $\dfrac{2}{4}$ 38. $\dfrac{30}{36}$

39. $\dfrac{98}{112}$ 40. $\dfrac{12}{108}$

41. $\dfrac{64}{656}$ 42. $\dfrac{16}{2048}$

43. $\dfrac{39}{65}$ 44. $\dfrac{119}{1700}$

45. $\dfrac{505}{1000}$

Copyright by The Goodheart-Willcox Co., Inc.

Unit 7
Addition of Fractions

Key Terms

common denominator   lowest common
denominator (LCD)

Introduction

The basic operations of addition, subtraction, multiplication, and division are performed with fractions and combinations of fractions and whole numbers. Handling these calculations is probably one of the most fundamental math jobs a welder is called upon to do.

Methods Used to Add Fractions

Three main situations will be encountered when adding fractions: fractions with common denominators, fractions without common denominators, and mixed numbers without common denominators.

Adding Fractions with Common Denominators

Fractions can be added only if they meet a specific requirement. They must have the <u>same</u> denominators. When two fractions have the same denominator, they share a **common denominator**. For example, $\frac{3}{11}$ and $\frac{4}{11}$ have a common denominator. To add these fractions, just add the numerators ($3 + 4 = 7$) and place this sum over the common denominator ($\frac{7}{11}$).

Adding Fractions without Common Denominators

Fractions without common denominators can be added by first changing the denominators so they are the same. The whole operation is easier if the denominator selected is the lowest one possible. Your first objective, then, is to find the **lowest common denominator (LCD)**. An efficient way of doing this is described below.

1. Review the fractions that are to be added.

$$\frac{5}{6} + \frac{3}{8} + \frac{2}{3}$$

2. In a blank space, write all the denominators in a row. It is a good idea to separate them by commas so they do not blend together.

$$6, 8, 3$$

3. Find a number that divides evenly into at least two of the numbers listed. In this example, 2 will work. Divide using the format illustrated. Record the results of the division.

$$2 \underline{|6, 8, 3}$$
$$3, 4, 3$$

Two divides evenly into 6 three times, so 3 is written below the 6. Two divides evenly into 8 four times, so 4 is written below the 8. Since 2 does not divide evenly into 3, the 3 is "brought down."

Copyright by The Goodheart-Willcox Co., Inc.

4. This step follows the pattern of the previous steps. This time, inspect the new line of 3, 4, 3 to find a number that divides evenly into at least two of the numbers. In this case, the number is 3.

$$\begin{array}{r} 2\,\big|\,6, 8, 3 \\ 3\,\big|\,3, 4, 3 \\ \hline 1, 4, 1 \end{array}$$

Since 3 does not divide evenly into 4, the 4 is brought down. At this point, the line of denominator digits cannot be reduced further.

5. Now multiply the line of reduced denominators (1, 4, 1) and the numbers by which they were divided (2, 3):

$$1 \times 4 \times 1 \times 2 \times 3 = 24$$

The LCD is 24.

6. Express each fraction from the original equation as an equivalent fraction with a denominator of 24. To review expressing equivalent fractions in higher terms, see Unit 6, *Introduction to Common Fractions*.

$$\frac{5}{6} + \frac{3}{8} + \frac{2}{3} \;=\; \frac{20}{24} + \frac{9}{24} + \frac{16}{24}$$

7. Add the numerators of the fractions.

$$\frac{20}{24} + \frac{9}{24} + \frac{16}{24} \;=\; \frac{45}{24}$$

8. If necessary, change to a mixed number and reduce the fraction to its lowest terms.

$$\frac{45}{24} \;=\; 1\frac{21}{24} \;=\; 1\frac{7}{8}$$

Adding Mixed Numbers without Common Denominators

There are several ways to add mixed numbers without common denominators. The most commonly used method is explained next.

1. Add $2\frac{3}{4} + 5\frac{1}{7}$

2. Rewrite the question in this manner.

$$\begin{array}{r} 2\frac{3}{4} \\ + \; 5\frac{1}{7} \\ \hline \end{array}$$

3. Find the LCD and write as equivalent fractions.

$$\begin{array}{r} 2\frac{21}{28} \\ + \; 5\frac{4}{28} \\ \hline \end{array}$$

Copyright by The Goodheart-Willcox Co., Inc.

4. Add the whole numbers together and then add the fractional parts together.

$$2 \frac{21}{28}$$

$$+ \quad 5 \frac{4}{28}$$

$$7 \frac{25}{28}$$

5. Reduce the answer to its lowest terms, if possible.
    Since $7^{25}/_{28}$ is already the lowest term in this case, no further operations are necessary.

Copyright by The Goodheart-Willcox Co., Inc.

Work Space

Copyright by The Goodheart-Willcox Co., Inc.

Name _____ Date _____ Class _____

Unit 7 Practice

Solve the following equations. Show all your work. Box your answers.

1. $\dfrac{3}{18} + \dfrac{7}{18} + \dfrac{1}{18} =$

2. $\dfrac{21}{96} + \dfrac{13}{96} + \dfrac{37}{96} =$

3. $\dfrac{109}{219} + \dfrac{25}{219} + \dfrac{34}{219} =$

4. $\dfrac{3}{5} + \dfrac{4}{15} =$

5. $\dfrac{17}{25} + \dfrac{4}{5} + \dfrac{49}{50} =$

6. $\dfrac{15}{28} + \dfrac{3}{4} + \dfrac{6}{7} =$

7. $\dfrac{3}{4} + \dfrac{1}{3} + \dfrac{4}{5} + \dfrac{1}{2} =$

8. $\dfrac{3}{8} + \dfrac{3}{7} + \dfrac{7}{12} =$

9. $\dfrac{107}{120} + \dfrac{85}{150} =$

10. $\dfrac{7}{27} + \dfrac{31}{36} + \dfrac{3}{23} =$

11. $\dfrac{7}{64} + \dfrac{41}{88} + \dfrac{53}{92} =$

12. $\dfrac{117}{365} + \dfrac{99}{200} + \dfrac{87}{260} =$

13. $13\dfrac{1}{4} + 5\dfrac{5}{8} + 4\dfrac{9}{16} =$

14. $\dfrac{3}{56} + \dfrac{3}{16} + \dfrac{7}{28} + \dfrac{2}{7} =$

15. $38\dfrac{19}{34} + 119\dfrac{11}{12} =$

16. $11\dfrac{1}{4} + 27\dfrac{3}{16} + \dfrac{7}{8} + 10\dfrac{1}{2} =$

17. $\dfrac{25}{140} + \dfrac{9}{204} =$

18. $1\dfrac{109}{187} + \dfrac{75}{121} =$

Copyright by The Goodheart-Willcox Co., Inc.

19. According to their time cards, the following people spent the indicated amount of time on Job #8815. What was the total amount of work time spent on the job?

Job #8815	
Name	Hours
Lee Jones	44¼
Jamil El Helou	185¾
Filomena Ferrari	121½
Pat Anderson	15
Luis Martin	29¼
Guy Laframbois	6½

20. A stainless steel rod, one-half inch in diameter, has been cut into five pieces. According to the lengths of each of the five pieces listed below, what was the original length of the rod? Ignore losses due to cutting.

> Twenty-one and three-quarter inches
>
> Seven inches
>
> Thirty-two and seven-thirty-seconds inches
>
> Eleven and one-half inches
>
> Sixteen and fifteen-sixteenth inches

21. At Camelot Metal Products, a welder has to cut a 1″ square rod into three pieces of the following lengths: 21⅜″, 26¼″, and 17⁵⁄₁₆″. Each cut will waste ¼″ of material. The pieces are cut from a 96″ rod. What length of the rod is used to make the three pieces?

22. What is the overall length of a machine part consisting of five pieces measuring ⅞″, 41⁹⁄₁₆″, 12²⁷⁄₃₂″, ¹³⁄₆₄″, and 1½″?

23. Leamington Brass produced a machined brass tube with an inside diameter of 2⅞″ and a thickness of ¹¹⁄₃₂″. What is the outside diameter?

24. The individual parts of a weldment had the following weights: 135½ lb, 19³⁄₁₆ lb, 71¼ lb, 6⅜ lb, and 1¼ lb. The filler material used in welding added an additional 2³⁄₃₂ lb. What is the final weight of the completed weldment?

Copyright by The Goodheart-Willcox Co., Inc.

Name _____   Date _____   Class _____

25. Find the distance from stud Ⓐ to stud Ⓑ?

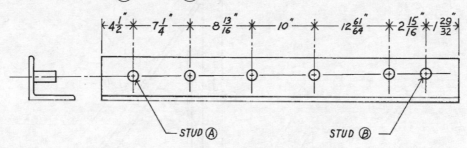

26. Calculate the overall length of the assembly in the diagram below.

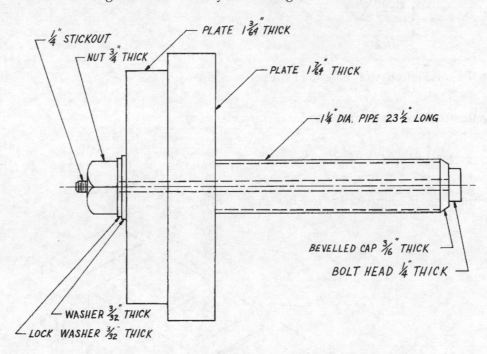

Copyright by The Goodheart-Willcox Co., Inc.

Review the measurements of the three objects in the diagram below to answer the following three questions.

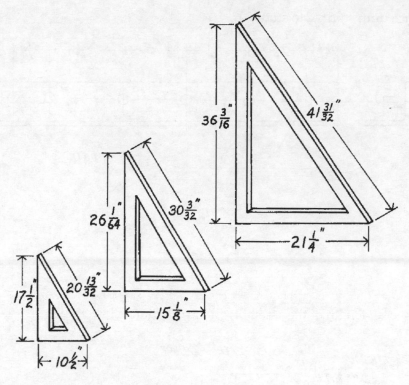

27. Calculate the total length of the vertical sides.

28. Calculate the total length of the horizontal sides.

29. Calculate the total length of the angular sides.

30. Calculate the overall length of this test bar after welding.

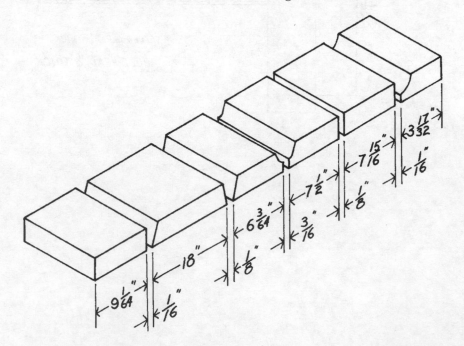

Copyright by The Goodheart-Willcox Co., Inc.

Name _____ Date _____ Class _____

Review the diagram below to answer the following two questions.

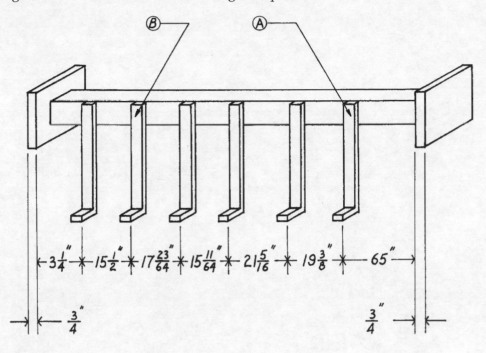

31. What is the distance from hanger Ⓐ to hanger Ⓑ?

32. What is the total length of the weldment above?

33. What is the total length of angle iron used in welding the following frame?

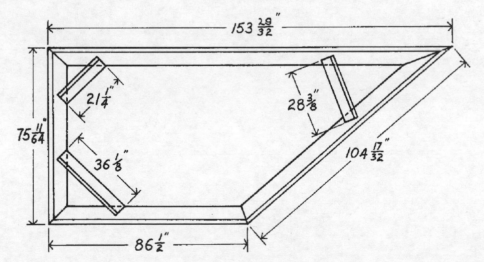

Copyright by The Goodheart-Willcox Co., Inc.

Work Space

Copyright by The Goodheart-Willcox Co., Inc.

Unit 8
Subtraction of Fractions

Introduction

Subtracting fractions is very similar to adding fractions. You will see the method is generally the same, except a subtraction, of course, is performed instead of an addition.

Methods Used to Subtract Fractions

Three main situations will be encountered where you will have to subtract fractions: fractions with common denominators, fractions without common denominators, and mixed numbers without common denominators.

Subtracting Fractions with Common Denominators

Fractions with common denominators are easily subtracted. You simply subtract one numerator from the other numerator and place the result over the common denominator.

$$\frac{9}{17} - \frac{3}{17} = \frac{6}{17}$$

Subtracting Fractions without Common Denominators

Fractions without common denominators can be subtracted by first changing the denominators so they are the same. As explained in Unit 7, *Addition of Fractions*, finding the LCD is usually the easiest way to do this. Review that explanation, if necessary. Then study the following example:

$$\frac{13}{14} - \frac{6}{63} =$$

$$7 \,|\, \underline{14, \, 63}$$
$$ 2, \;\; 9$$

$$7 \times 2 \times 9 = 126$$

$$\text{LCD} = 126$$

Therefore:

$$\frac{13}{14} - \frac{6}{63} = \frac{117}{126} - \frac{12}{126} = \frac{105}{126}$$

Copyright by The Goodheart-Willcox Co., Inc.

Subtracting Mixed Numbers without Common Denominators

To subtract mixed numbers without common denominators, first change the fractional parts to common denominators. Then subtract the fractional parts. Finally, subtract the whole numbers.

$$17\,\frac{3}{4} = 17\,\frac{21}{28}$$

$$-4\,\frac{1}{7} = -4\,\frac{4}{28}$$

$$\overline{\qquad\qquad 13\,\frac{17}{28}}$$

Borrowing a Whole Number for a Fraction

Not all subtraction equations follow the exact steps listed above. A complication occurs when the fractional part to be subtracted (the subtrahend) is larger than the fraction from which it is being subtracted (the minuend).

$$11\,\frac{9}{16}$$

$$-5\,\frac{12}{16}$$

$$\overline{}$$

When this situation occurs, borrow 1 from the whole number and add it to the fraction part of the number.

$$11\!\!\!\!{}^{\,0}\,\frac{9}{16}\left(+\frac{16}{16}\right)$$

$$-5\,\frac{12}{16}$$

$$\overline{}$$

When a fraction borrows from a whole number, the whole number must be changed into a fraction before it can be added to fraction. When a 1 is changed to a fraction, it has the same number in the numerator as it has in the denominator. Examples of how the number 1 appears as a fraction is as follow: ⅓, ⅝, ¹⁶/₁₆. For review, see Unit 6, *Introduction to Common Fractions*.

$$10\,\frac{25}{16}$$

$$-5\,\frac{12}{16}$$

$$\overline{\qquad\qquad 5\,\frac{13}{16}}$$

After reducing the whole number from the borrowing and calculating the new fractional numerator, the problem can be solved by subtracting each column right to left.

Copyright by The Goodheart-Willcox Co., Inc.

Name _____ Date _____ Class _____

Unit 8 Practice

Solve the following equations. Reduce to lowest terms. Show all your work. Box your answers.

1. $\dfrac{5}{9} - \dfrac{2}{9} =$

2. $\dfrac{13}{16} - \dfrac{5}{16} =$

3. $\dfrac{62}{101} - \dfrac{31}{101} =$

4. $\dfrac{7}{8} - \dfrac{1}{4} =$

5. $\dfrac{15}{16} - \dfrac{7}{8} =$

6. $35\dfrac{17}{20} - 25\dfrac{11}{17} =$

7. Reduce $738\frac{7}{32}$ by $48\frac{9}{32}$.

8. Find the difference between $6{,}499\frac{3}{4}$ and $6{,}460\frac{1}{2}$.

9. $111\frac{1}{119}$ less $111\frac{1}{238}$.

10. Subtract $\frac{19}{64}$ from $\frac{5}{8}$.

11. Take $49\frac{2}{3}$ away from 65.

12. Calculate the difference between $2{,}020\frac{1}{2}$ and $1{,}100\frac{1}{2}$.

13. $7\dfrac{1}{3} - 6\dfrac{2}{3} =$

14. $11\dfrac{9}{16} - \dfrac{7}{8} =$

15. $1\dfrac{3}{10} - \dfrac{5}{6} =$

16. $19\dfrac{2}{5} - 18\dfrac{9}{10} =$

17. $45\dfrac{1}{3} - 32\dfrac{2}{3} =$

18. $24\dfrac{1}{4} - 15\dfrac{2}{3} =$

Copyright by The Goodheart-Willcox Co., Inc.

19. A piece of steel (U.S. Standard gage #28), 4′ and 7⅝″ long was sheared from a sheet 9′ and 10⁷⁄₁₆″ long. What length of the original sheet remains?

20. The sides of a square piece of steel measure 18⅜″. Using two cuts, the piece is sheared to 11⁷⁄₆₄″ × 13³⁄₁₆″. What is the width of each removed piece?

Use this information to answer the three questions that follow. A length of Minnesota pipeline measuring 87½ miles is due for inspection. The contract went to Twin City Consulting, Inc. The company estimated they could inspect 5⅛ miles of line per week.

21. What length of pipeline would remain to be inspected after one week?

22. What length of pipeline would remain to be inspected after two weeks?

23. What length of pipeline would remain to be inspected after three weeks?

24. In Detroit's Industrial Softball League, halfway through the season the Pipefitters were 17 games behind the leader and ranked at 5th place. The Boilermakers were in 4th place and 13½ games behind the leader. How many games behind the Boilermakers were the Pipefitters?

25. The wall thickness of a piece of round tubing is ³⁄₁₆″ and the outside diameter is 3¼″. What is the inside diameter?

26. Four pieces of square tubing measuring 22⁵⁄₃₂″, 17″, 20½″, and 21¼″ are to be cut from a stock piece of 1″ × 1″ square tubing that is 131¼″ long. Each saw cut wastes ³⁄₃₂″ of material. What will be the length of the stock piece after cutting all four pieces?

27. If the ¼″ square bar below is reduced to 20¼″, how long would the piece removed be? The saw cut will waste ¹⁄₁₆″ of material.

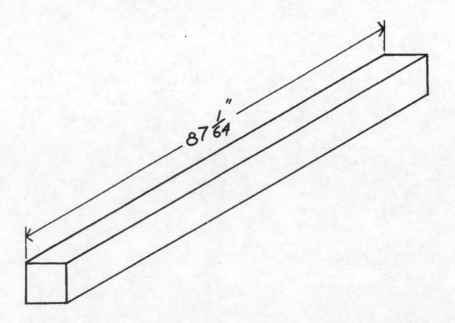

Copyright by The Goodheart-Willcox Co., Inc.

Name _____ Date _____ Class _____

28. If the 1¼″ × 1¼″ square tubing below is reduced to 23⅝″, how long would the piece removed be? The saw cut will waste ¹⁄₁₆″ of material.

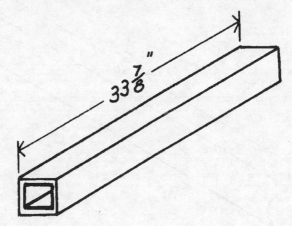

29. If the 2″ channel below is reduced to 40⅝″, how long would the piece removed be? The saw cut will waste ¹⁄₁₆″ of material.

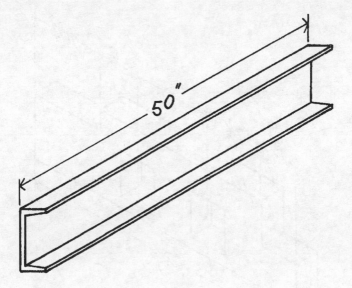

Copyright by The Goodheart-Willcox Co., Inc.

30. If the ¾" round bar below is reduced to 38¼", how long would the piece removed be? The saw cut will waste ¹⁄₁₆" of material.

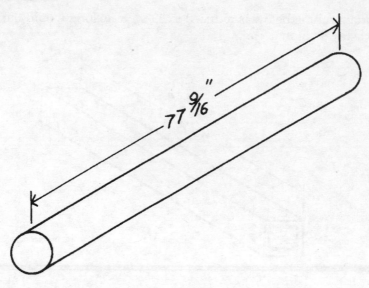

31. In the diagram below, calculate the length of Ⓐ.

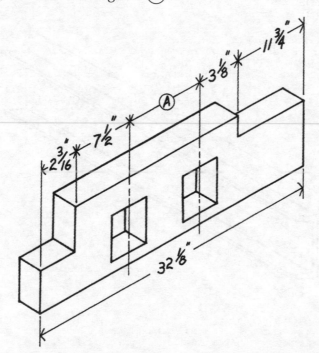

Copyright by The Goodheart-Willcox Co., Inc.

Name _____ Date _____ Class _____

Refer to the diagram below to answer the questions that follow.

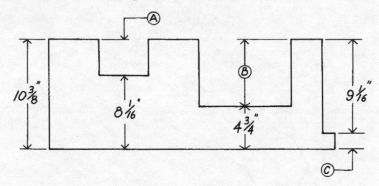

32. In the diagram above, find the length of (A).

33. In the diagram above, find the length of (B).

34. In the diagram above, find the length of (C).

35. Slots are cut from the circular piece below. Calculate the length of (A).

36. Calculate the length of (B) in the diagram of the slotted, circular piece.

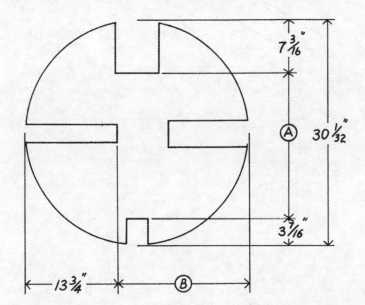

Copyright by The Goodheart-Willcox Co., Inc.

37. Find the length of Ⓐ in the diagram below.

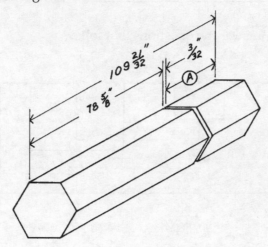

38. Three pieces measuring 18¹¹⁄₆₄″, 9⅜″, and 82¹⁄₁₆″ will be cut from this half-round stock. Each saw cut wastes ³⁄₃₂″ of material. What will be the length of the remaining piece?

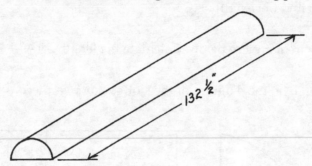

Copyright by The Goodheart-Willcox Co., Inc.

Name _____ Date _____ Class _____

Review the diagram below to answer the questions that follow.

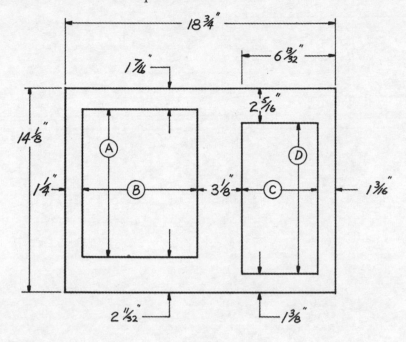

39. Calculate the length of (A).

40. Calculate the length of (B).

41. Calculate the length of (C).

42. Calculate the length of (D).

Copyright by The Goodheart-Willcox Co., Inc.

Work Space

Copyright by The Goodheart-Willcox Co., Inc.

Multiplication of Fractions

Introduction

It may seem surprising, but multiplying fractions is easier than adding or subtracting fractions. This is because the denominators do not have to be the same. You do not have to find a common denominator.

Three expressions are commonly used to indicate multiplication of fractions.

$$\frac{1}{5} \times \frac{3}{4}$$

$$\frac{1}{5} \text{ times } \frac{3}{4}$$

$$\frac{1}{5} \text{ of } \frac{3}{4}$$

The one expression that may seem peculiar is *of*; however, through common usage, this word has come to mean multiplication.

Methods Used to Multiply Fractions

To multiply ¾ × ⁵⁄₇, multiply the two numerators and then multiply the two denominators.

$$\frac{3}{4} \times \frac{5}{7} = \frac{15}{28}$$

As always, if the answer can be reduced to lower terms, you would do so. Sometimes a question can be reduced to a simpler form before beginning the multiplication. You can do this by finding a number that will divide evenly into *any* of the numerators and denominators. To better understand this, you may want to rewrite the equation as a visual cue.

$$\frac{16}{100} \times \frac{25}{37} = \frac{16 \times 25}{100 \times 37}$$

Arranging a fraction multiplication equation as a single fraction with one numerator and one denominator should remind you of the flexibility allowed in reducing the equation. Now, begin reducing the terms.

$$\frac{16 \times \overset{1}{\cancel{25}}}{\underset{4}{\cancel{100}} \times 37} =$$

Always look to reduce to the lowest terms possible. Even after reducing the 25 and 100, this equation can be reduced further.

$$\frac{\overset{4}{\cancel{16}} \times \overset{1}{\cancel{25}}}{\underset{\underset{1}{\cancel{4}}}{\cancel{100}} \times 37} =$$

Copyright by The Goodheart-Willcox Co., Inc.

Continue doing this until the numerators and denominators cannot be reduced any further. Then perform the multiplication

$$\frac{\overset{4}{\cancel{16}} \times \overset{1}{\cancel{25}}}{\underset{4}{\cancel{100}} \times 37} = \frac{4}{37}$$

You will find that multiplying fractions is much easier if you first reduce to the lowest possible terms.

Multiplying Mixed Numbers

To multiply mixed numbers, first change them to improper fractions. Once the multiplication is done, convert the product back into a mixed number, as shown here.

$$3\frac{1}{2} \times 9\frac{2}{5} = \frac{7}{2} \times \frac{47}{5} = \frac{329}{10} = 32\frac{9}{10}$$

If the task is to multiply a mixed number and a whole number, rewrite the whole number as a fraction with a denominator of 1. For example, 42 is rewritten as $^{42}/_1$. A typical problem would look like this:

$$42 \times 2\frac{4}{7} =$$

$$\frac{42}{1} \times \frac{18}{7} =$$

$$\frac{\overset{6}{\cancel{42}} \times 18}{1 \times \underset{1}{\cancel{7}}} =$$

$$\frac{6 \times 18}{1 \times 1} = \frac{108}{1} = 108$$

Multiplying More than Two Fractions

Multiplying more than two fractions follows the same routine as multiplying two fractions. First, try to reduce the question to a simpler form and then multiply the numerators and denominators.

$$\frac{1}{2} \times \frac{\overset{1}{\cancel{3}}}{4} \times \frac{5}{\underset{2}{\cancel{6}}}$$

$$\frac{1}{2} \times \frac{1}{4} \times \frac{5}{2} = \frac{5}{16}$$

Copyright by The Goodheart-Willcox Co., Inc.

Multiplying More than Two Mixed Numbers

When an equation is written to multiply more than two mixed numbers, first convert the mixed numbers into improper fractions.

$$5\frac{1}{2} \times 4\frac{2}{3} \times 9\frac{3}{5} = \frac{11}{2} \times \frac{14}{3} \times \frac{48}{5}$$

Next, compare numerators and denominators. Reduce any fractions to a simpler equivalent form.

$$\frac{11}{\overset{2}{\underset{1}{\cancel{2}}}} \times \frac{\overset{7}{\cancel{14}}}{\underset{1}{\cancel{3}}} \times \frac{\overset{16}{\cancel{48}}}{5}$$

Now calculate the equation by multiplying the numerators. Also, multiply the denominators together.

$$\frac{11}{1} \times \frac{7}{1} \times \frac{16}{5} = \frac{1,232}{5}$$

The resulting improper fraction can be turned into a mixed number by dividing the numerator by the denominator. The quotient becomes the whole number. The divisor becomes the denominator. The remainder becomes the numerator.

$$
\begin{array}{r}
246 \text{ r } 2 = 246\frac{2}{5} \\[4pt]
5\overline{)1{,}232} \\
-10\downarrow \\
\overline{23} \\
-20\downarrow \\
\overline{32} \\
-30 \\
\overline{2}
\end{array}
$$

Copyright by The Goodheart-Willcox Co., Inc.

Work Space

Copyright by The Goodheart-Willcox Co., Inc.

Name _____ Date _____ Class _____

Unit 9 Practice

Solve the following equations. Show all your work. Box your answers.

1. $\dfrac{1}{5} \times \dfrac{1}{3} =$

2. $\dfrac{3}{4} \times \dfrac{7}{15} =$

3. $\dfrac{15}{33} \times \dfrac{12}{61} =$

4. $\dfrac{1}{2} \times \dfrac{1}{2} =$

5. $\dfrac{13}{64} \times \dfrac{8}{9} =$

6. $\dfrac{7}{35} \times \dfrac{12}{144} =$

7. $4 \times \dfrac{3}{16} =$

8. $3\dfrac{1}{3} \times 5\dfrac{1}{2} =$

9. $17\dfrac{2}{5} \times 2\dfrac{1}{4} =$

10. $8\dfrac{1}{2} \times \dfrac{15}{45} =$

11. $9\dfrac{3}{8} \times 9\dfrac{3}{8} =$

12. $15 \times \dfrac{15}{16} =$

13. $\dfrac{16}{27} \times \dfrac{3}{4} \times \dfrac{5}{7} =$

14. $3\dfrac{1}{2} \times 4\dfrac{1}{6} \times 1\dfrac{3}{10} =$

15. $11\dfrac{3}{16} \times 1\dfrac{1}{8} \times 2 =$

16. $\dfrac{23}{140} \times \dfrac{70}{92} \times \dfrac{17}{33} =$

17. $7 \times \dfrac{1}{2} \times \dfrac{3}{4} =$

18. $9 \times \dfrac{4}{15} \times 9\dfrac{4}{15} =$

Copyright by The Goodheart-Willcox Co., Inc.

19. A pipeline is laid at the rate of ⅛ mile per day. How many miles of line would be completed in 29½ days?

20. One cubic foot of water weighs 62½ lb. Find the weight of the contents of a welded steel tank containing 28 cubic feet of water.

21. Last year, Solar Supplies lost 133½ man hours of labor due to accidents. This year, they lost 1½ times that amount. How many hours were lost this year?

22. Kilowatts are changed to horsepower by multiplying the number of kilowatts by 1⅓. Change 24½ kilowatts to horsepower.

23. Average truck drivers can move their foot from the accelerator to the brake in ¹³⁄₁₆ of a second. At 60 miles per hour a truck is traveling 88 feet per second. How far will the truck travel before the driver's foot is moved to the brake?

Use this information to answer the two questions that follow. Three years ago, YKY Fasteners International produced 5,960,000 bolts of various sizes. The production two years ago was 1⅗ times the total from three years ago. Last year's production was 1¹³⁄₁₆ times the total from three years ago.

24. How many bolts were produced two years ago?

25. How many bolts were produced last year?

26. Two grain hoppers were built by a crew consisting of three welders who worked on the job 4¼ hours a day for 17 days. How many hours did the crew work on the hoppers?

27. A GMAW welder running at a speed of 16⅜" per minute requires 14½ minutes to complete a V-groove weld on an aluminum sill. What is the length of the sill?

Use this information to answer the three questions that follow. During each hour, a shop uses the following quantities of weld material:

    132⅜ lb of ³⁄₁₆" wire

    81⁵⁄₁₆ lb of ⁵⁄₆₄" wire

    25¾ lb of ¼" wire

28. How many pounds of ¼" wire are used in 5⅙ hours?

29. How many pounds of ³⁄₁₆" wire are used in 8½ hours?

30. How many pounds of ⁵⁄₆₄" wire are used in 3⁷⁄₆₀ hours?

Copyright by The Goodheart-Willcox Co., Inc.

Name _____ Date _____ Class _____

31. The following weldment requires 6⅔ rods. If 40½ weldments are completed in 2¼ days, how many rods will be required?

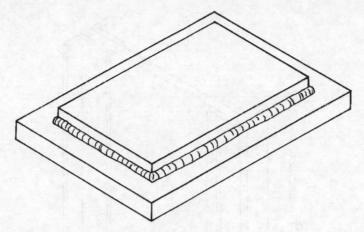

32. Review the diagram below. All notches are 2¹¹⁄₁₆″ wide. All contact surfaces are 3⅜″ wide. What is the overall length of the rack?

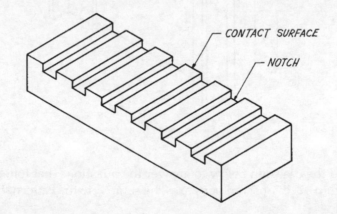

Copyright by The Goodheart-Willcox Co., Inc.

33. Review the diagram below. Nine of these pillars are to be fabricated. For the purpose of weight reduction, each pillar is to have 13 holes flame cut in the center plate. Seven large holes are to be cut, reducing the weight by $32\frac{19}{32}$ lb per hole. Six smaller holes are to be cut, reducing the weight by $23\frac{7}{16}$ lb per hole. What is the total weight reduction for the entire project?

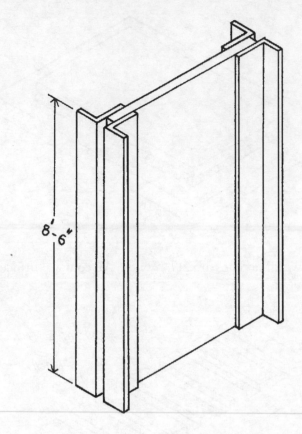

Use this information and the diagram below to answer the questions that follow. A flame cutting job using Pattern #1 resulted in $\frac{3}{16}$ lb of scrap per part. The scrap rate for Pattern #2 was $\frac{2}{5}$ of the rate for Pattern #1.

34. What weight of scrap would result if Pattern #1 is used to produce 5,040 parts?

35. What weight of scrap would result if Pattern #2 is used to produce the same number of parts?

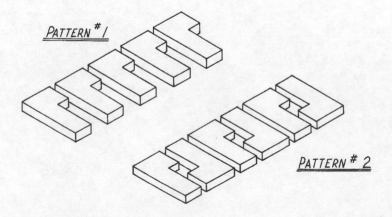

Copyright by The Goodheart-Willcox Co., Inc.

Name _____ Date _____ Class _____

Use this information and the diagram below to answer the two questions that follow. One part, as illustrated below, is produced every $^{47}/_{60}$ of a minute. One weld bead is used to join each vertical length of the top piece to the lower piece.

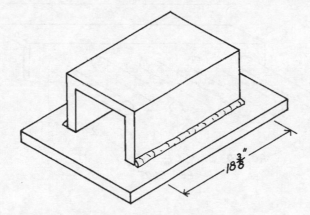

36. How many minutes are required to produce 2,864 parts?

37. What is the total length of weld deposited for 2,864 parts?

Use this information and the diagram below to answer the two questions that follow. The structural work for a new shopping mall includes 1,047 hangers.

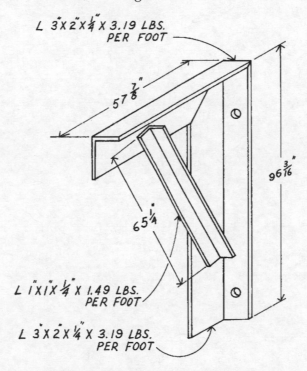

38. What is the total length of 3″ × 2″ angle used?

39. What is the total length of 1″ × 1″ angle used?

Copyright by The Goodheart-Willcox Co., Inc.

40. Refer to the diagram below. Calculate the weight of the weldment.

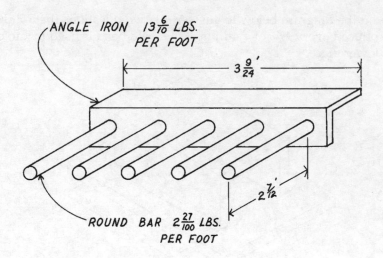

Copyright by The Goodheart-Willcox Co., Inc.

Unit 10
Division of Fractions

Key Terms

complex fraction   invert the divisor

Introduction

Learning to divide fractions is quite easy once you have learned to multiply fractions. You will see why as you read this unit.

Methods Used to Divide Fractions

Fractional division equations closely resemble fractional multiplication equations. In both cases, the fractions are written vertically and separated by the operational math symbol. To divide fractions, **invert the divisor** (that is, turn the divisor upside down by switching the numerator and denominator) and change the operation symbol from division to multiplication.

$$\frac{3}{5} \div \frac{5}{8} = \frac{3}{5} \times \frac{8}{5} = \frac{24}{25}$$

Remember, the divisor is the number doing the division. In this example, ⅗ (the dividend) is being divided by ⅝, the divisor. When dividing fractions, be certain to identify the divisor. If the wrong fraction is inverted, your answer will most likely be incorrect.

It may seem odd that a division problem can suddenly be switched to a multiplication problem. Although they are opposite operations, multiplication and division are closely related. Because of this close relationship, it is possible to convert fractional division problems to multiplications as shown above.

Dividing Mixed Numbers

As in multiplying fractions, first change the mixed numbers to improper fractions. Then, invert the divisor and multiply. As usual, try to reduce the problem to a simpler form before proceeding with the multiplication part of the problem. Here is an example:

$$7\frac{3}{4} \div 7\frac{1}{2} =$$

$$\frac{31}{4} \div \frac{15}{2} =$$

$$\frac{31}{\overset{4}{\underset{2}{}}} \times \frac{\overset{1}{\cancel{2}}}{15} = \frac{31}{30} = 1\frac{1}{30}$$

Copyright by The Goodheart-Willcox Co., Inc.

Dividing Complex Fractions

There are several ways to symbolize division of fractions. Here are two different formats you can expect to encounter:

The standard format:

$$\frac{5}{8} \div \frac{2}{3}$$

The complex fraction format:

$$\frac{\frac{5}{8}}{\frac{2}{3}}$$

A **complex fraction** has a fraction for the numerator and a fraction for the denominator. To easily perform this division, rewrite it in the standard format and proceed in the usual manner:

$$\frac{\frac{5}{8}}{\frac{2}{3}} = \frac{5}{8} \div \frac{2}{3} = \frac{5}{8} \times \frac{3}{2} = \frac{15}{16}$$

So, you can see that even though a complex fraction may appear at first very difficult to divide, it really is quite simple. The same general procedure is followed when the numerator and/or denominator contain mixed numbers. However, in such cases, the mixed number is converted into an improper fraction.

$$\frac{\frac{9}{144}}{6\frac{3}{4}} = \frac{9}{144} \div 6\frac{3}{4} = \frac{9}{144} \div \frac{27}{4} = \frac{9}{144} \times \frac{4}{27} = \frac{\overset{1}{\cancel{9}}}{\underset{36}{\cancel{144}}} \times \frac{\overset{1}{\cancel{4}}}{\underset{3}{\cancel{27}}} = \frac{1}{108}$$

Copyright by The Goodheart-Willcox Co., Inc.

Name _____ Date _____ Class _____

Unit 10 Practice

Solve the following equations. Show all your work. Box your answers.

1. $\dfrac{1}{3} \div \dfrac{5}{16} =$

2. $\dfrac{4}{5} \div \dfrac{9}{10} =$

3. $\dfrac{5}{6} \div \dfrac{7}{8} =$

4. $\dfrac{7}{11} \div \dfrac{7}{11} =$

5. $\dfrac{3}{4} \div \dfrac{4}{3} =$

6. $\dfrac{1}{2} \div \dfrac{1}{4} =$

7. $21 \div \dfrac{5}{6} =$

8. $2\dfrac{5}{16} \div 13 =$

9. $57 \div 2\dfrac{5}{8} =$

10. $63 \div \dfrac{1}{63} =$

11. $9\dfrac{3}{16} \div 3\dfrac{9}{16} =$

12. $100 \div \dfrac{1}{2} =$

13. $1\dfrac{13}{16} \div 2\dfrac{1}{4} =$

14. $5\dfrac{4}{9} \div 5\dfrac{4}{9} =$

15. $\dfrac{7\dfrac{9}{16}}{7} =$

16. $\dfrac{4\dfrac{2}{5}}{2\dfrac{3}{4}} =$

Copyright by The Goodheart-Willcox Co., Inc.

17. $\dfrac{3\frac{5}{6}}{9\frac{1}{5}}$ =

18. $\dfrac{\frac{123}{716}}{\frac{82}{182}}$ =

19. How many 3½″ pieces can be sheared from a thin piece of sheet metal 31″ long?

20. A piece of ½″ diameter copper tubing 105½″ long is cut with a pipe cutter into 6 pieces of equal length. What is the length of each piece?

Use this information to answer the two questions that follow. A bus, rented by E & P Mold Co., is to deliver a crew of workers to a job site 87½ miles away. The bus departs from the shop at 6:30 a.m. and is expected to arrive at 9:00 a.m.

21. What average speed must the bus maintain to arrive on schedule?

22. If the bus averages 19½ miles per gallon, how many gallons of fuel are needed for one trip to the site and back to the shop?

23. A sheet of metal weighs 18⁷⁄₁₆ lb. In a shearing operation, the sheet is cut into strips weighing ⅜ lb each. How many strips of metal are produced?

24. A steel plate 10′ 6½″ long weighs 366¹⁷⁄₂₀ lb. How much does a 1″ length of the plate weigh?

25. A volume of 1 cubic foot contains about 7½ gallons. How many cubic feet of oil will a cooling tank hold if it contains 1,776½ gallons?

26. Divide this T-shape into 13 equal parts. What length are the pieces?

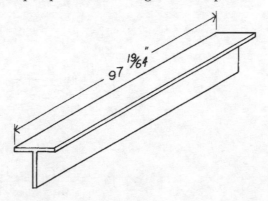

Copyright by The Goodheart-Willcox Co., Inc.

Name _____ Date _____ Class _____

27. The following square bar of hot rolled steel is cut into four equal length pieces. Each saw cut wastes $\frac{3}{32}''$ of material. What length are the pieces?

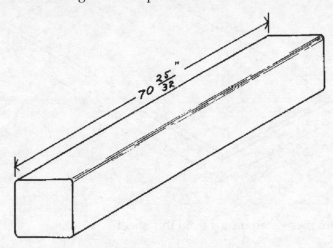

28. The cross section of a piece of extra strong pipe has the following dimensions. What is the wall thickness of the pipe?

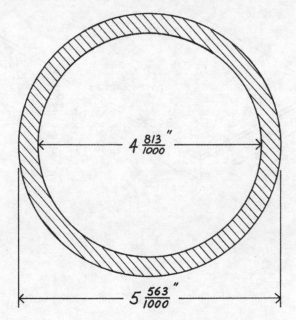

Copyright by The Goodheart-Willcox Co., Inc.

29. How many strips 48″ long and 13⁷⁄₁₆″ wide can be sheared from this sheet of 15 gage steel?

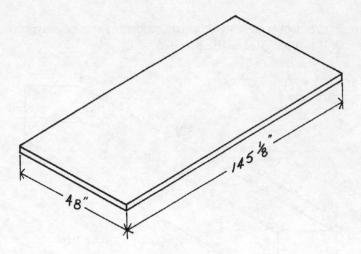

30. How many 7¾″ square pieces can be cut from this sheet?

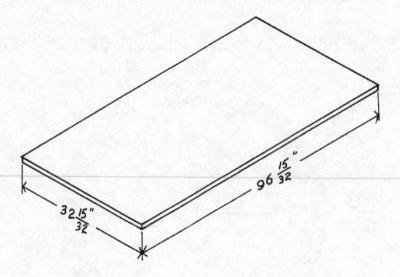

Copyright by The Goodheart-Willcox Co., Inc.

Name _____ Date _____ Class _____

31. In the diagram below, fifteen more pieces of square tubing will be welded to the top side of this plate. Each tube will be spaced an equal distance from the next tube. Calculate the distance between any two consecutive tube pieces.

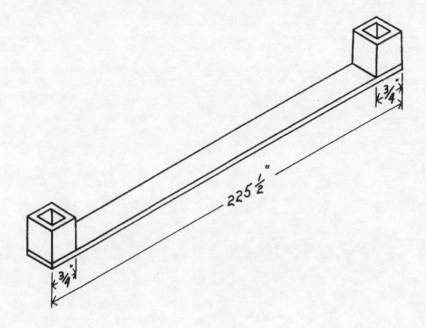

32. Three more rods of 1¼″ diameter and four more rods of 1¾″ diameter are to be welded to this plate. Center lines are to be equally spaced. What is the distance between centers?

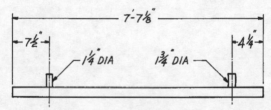

Copyright by The Goodheart-Willcox Co., Inc.

Work Space

Copyright by The Goodheart-Willcox Co., Inc.

Section 3
Decimal Fractions

Section Objectives

After studying this section, you will be able to:

- Give examples of decimal fractions
- Explain how to convert decimal fractions and common fractions
- Show how to round decimal fractions
- Perform addition of decimal fractions
- Perform subtraction of decimal fractions
- Perform multiplication of decimal fractions
- Perform division of decimal fractions

Copyright by The Goodheart-Willcox Co., Inc.

Unit 11
Introduction to Decimal Fractions

Key Terms

decimal fraction   rounding the decimal
decimal point

Introduction

As you have seen so far, numbers can be classified in a variety of ways, such as whole numbers, common fractions, and mixed numbers. This unit introduces another classification of numbers called decimal numbers.

In Unit One, it was noted that our numbering system is based on ten digits and that larger numbers are created by lining up these digits in a certain order. The decimal system uses this method of place position values to express numbers that are *less* than whole numbers. These are called **decimal fractions**.

All decimal numbers include a **decimal point**. Digits to the left of the decimal point are whole numbers. Digits to the right of the decimal point are fractional numbers. Each position to the right of the decimal has a value. Listed below are some of the names and place values.

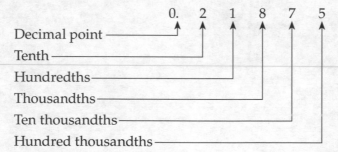

There are two ways of expressing decimal fractions verbally:

- Pronounce each digit individually, such as "14.63" as "Fourteen point (or decimal) six three."

- Pronounce the decimal fraction as a whole number and add the name of the last place value, such as "14.63" as "Fourteen and sixty-three hundredths."

Always include a zero to the left of the decimal point if there are no whole numbers in the units place.

Converting Decimal Fractions to Common Fractions

A decimal fraction is changed to a common fraction by using the last place value as the denominator and the decimal fraction digits as the numerator.

$$0.3 = \frac{3}{10}$$

$$0.27 = \frac{27}{100}$$

Copyright by The Goodheart-Willcox Co., Inc.

$$0.173 = \frac{173}{1000}$$

The task is simplified by the fact that denominators will always be multiples of 10, such as 10, 100, 1,000, 10,000, etc. The actual value depends upon the place position of the last digit. Once you have done the conversion, reduce the fraction, if possible.

$$0.0625 = \frac{625}{10,000} = \frac{25}{400} = \frac{1}{16}$$

$$0.828125 = \frac{828,125}{1,000,000} = \frac{33,125}{40,000} = \frac{53}{64}$$

Converting Common Fractions to Decimal Fractions

A common fraction is changed to a decimal fraction by dividing the numerator by the denominator:

$$\frac{3}{4} = 4\overline{)3}$$

Place a decimal to the right of the dividend. Place another decimal directly above that decimal. This is done so the decimal in the answer (the quotient) is properly located:

$$\frac{3}{4} = 4\overline{)3.}$$

Now, add a zero to the right of the decimal in the dividend and begin to divide. Add zeros to the dividend as needed:

$$\frac{3}{4} = \begin{array}{r} .75 \\ 4\overline{)3.00} \\ -2\,8\downarrow \\ \hline 20 \\ -20 \\ \hline 0r \end{array}$$

Rounding Decimals

The previous example is typical of most fractions. However, there are two situations you should watch for and take special action. First, there are some common fractions that result in an unending decimal fraction. For example, ⅓ will convert to the unending decimal fraction 0.33333. Second, some common fractions will convert to decimal fractions that are quite long and much more accurate than the original common fraction. Therefore, it is common practice to reduce such numbers to a degree of accuracy that is adequate. This process is called **rounding the decimal** and is done as follows:

Determine the degree of accuracy required. Normally, the problem description will indicate the degree of accuracy required or to what place the decimal fraction should be extended. Also, the accuracy required is usually indicated as a tolerance on the shop drawings you will be using. An explanation of tolerance will be covered later in this text.

Copyright by The Goodheart-Willcox Co., Inc.

Eliminate all digits beyond the required degree of accuracy. If the first number you eliminate is 5 or more, increase the final number in your answer by 1. Following is an example:

1. Round 75.13846 to the nearest hundredth.

2. 75.13~~846~~

3. Since 8, the first number eliminated, is greater than 5, increase the 3 to a 4.

4. The answer is 75.14.

Copyright by The Goodheart-Willcox Co., Inc.

Name _____   Date _____   Class _____

Unit 11 Practice

Write the following as decimal numbers.

1. One hundred decimal zero one:

2. Ninety five hundredths:

3. Fourteen decimal zero zero one two five:

4. Three thousand two hundred nineteen decimal one two five:

5. Decimal seven zero seven:

6. One thousand nine hundred seventeen ten thousandths:

7. Decimal eight six six:

8. Five decimal five one five six three:

Convert the following decimal fractions to common fractions. Box your answers.

9. 0.45   10. 0.28125

11. 0.1335   12. 0.6875

13. 0.3333   14. 0.78125

Convert the following to decimal fractions. Box your answers.

15. $\dfrac{1}{2}$ 16. $100\dfrac{1}{100}$

17. $\dfrac{1}{32}$ 18. $\dfrac{45}{64}$

19. $\dfrac{5}{8}$ 20. $\dfrac{3}{1,000}$

Copyright by The Goodheart-Willcox Co., Inc.

Round the following decimals to the nearest tenth. Box your answers.

21. 0.8801   22. 0.1718

23. 0.4691   24. 0.5555

25. 0.0923   26. 0.0505

Round the following decimals to the nearest hundredth. Box your answers.

27. 0.9551   28. .02192

29. 0.0916   30. 0.0747

31. 0.9097   32. 0.9999

Round the following decimals to the nearest thousandth. Box your answers.

33. 0.3138   34. 0.09001

35. 0.0091   36. 0.4375

37. 0.4426   38. 0.0541

Convert the following common fractions to decimal fractions and round to the nearest hundredth.

39. $\dfrac{2}{3}$ 40. $\dfrac{3}{4}$

41. $\dfrac{1,111}{10,000}$ 42. $\dfrac{987}{1,000}$

43. $\dfrac{1,001}{10,000}$ 44. $\dfrac{23}{43}$

Copyright by The Goodheart-Willcox Co., Inc.

Name _____ Date _____ Class _____

Convert the following common fractions to decimal fractions and round to the nearest thousandth.

45. $\dfrac{15}{16}$ 46. $\dfrac{3}{7}$

47. $\dfrac{1}{16}$ 48. $\dfrac{21}{43}$

49. $\dfrac{7,666}{10,000}$ 50. $\dfrac{75,196}{100,000}$

Convert the following common fractions to decimal fractions and round to the nearest ten thousandth.

51. $\dfrac{1}{64}$ 52. $\dfrac{13}{15}$

53. $\dfrac{12,345}{100,000}$ 54. $\dfrac{1}{9}$

55. $\dfrac{198}{205}$ 56. $\dfrac{31}{32}$

Copyright by The Goodheart-Willcox Co., Inc.

Work Space

Copyright by The Goodheart-Willcox Co., Inc.

Unit 12
Addition and Subtraction of Decimal Fractions

Introduction

Addition and subtraction of decimal fractions are explained together in this unit because the two operations share important characteristics.

Method Used to Add Decimal Fractions

To add decimal numbers, write them in a column with the decimal points lined up, as follows:

$$
\begin{array}{r}
7.2 \\
19.01 \\
.6 \\
+\ 100.407 \\
\end{array}
$$

Be sure the decimal points are accurately lined up. You may find it helpful to add zeros so the far right side of the column is filled. Add all the digits starting from the right as if they are all whole numbers. Carry over any numbers necessary into the next column.

$$
\begin{array}{r}
11 \\
7.200 \\
19.010 \\
.600 \\
+\ 100.407 \\
\hline
127.217 \\
\end{array}
$$

Method Used to Subtract Decimal Fractions

To subtract decimal fractions, you must also line up the decimal points. As in addition, you may add zeroes to the far right side of numbers to fill the column and maintain decimal point alignment. Subtract each column starting from the right as if they are all whole numbers. When necessary, borrow ones from the column to the left.

$$
\begin{array}{r}
8\,1 \\
38.947 \\
-\ 13.760 \\
\hline
25.187 \\
\end{array}
$$

The operation is the same as that for whole numbers except for the decimal point. The decimal place values are treated just the same as the values to the left of the decimal point. Review Unit Three, *Subtraction of Whole Numbers*, to brush up on carrying to fill place values.

Copyright by The Goodheart-Willcox Co., Inc.

Work Space

Copyright by The Goodheart-Willcox Co., Inc.

Name _____ Date _____ Class _____

Unit 12 Practice

Perform the following equations. Show all your work. Be certain the columns line up. Box your answers.

1. $0.3 + 9.2 + 3.9 =$

2. $2.08 + 4.160 + 0.69 =$

3. $0.354 + 354.009 + 9.035 =$

4.
$$
\begin{array}{r}
3.41 \\
2.45 \\
4.67 \\
+\ 5.26 \\
\hline
\end{array}
$$

5.
$$
\begin{array}{r}
0.00532 \\
0.138 \\
4.322 \\
+\ 48.2 \\
\hline
\end{array}
$$

6.
$$
\begin{array}{r}
0.109 \\
69.96 \\
10.01 \\
+\ 0.2 \\
\hline
\end{array}
$$

7. $16.601 + 0.195 + 4.749 + 0.945 + 200 =$

8. $0.049 + 1{,}741.9 + 33.0 + 0.1 + 3.2 =$

9. $0.6593 + 0.4978 + 100.0 + 3.1416 + 0.0101 =$

10. $0.183 + 0.222 + 0.296 + 0.252 + 0.243 + 4.801 + 0.395 + 2.424 + 1.573 + 0.244 + 0.391 =$

11.
$$
\begin{array}{r}
0.9 \\
-\ 0.2 \\
\hline
\end{array}
$$

12.
$$
\begin{array}{r}
19.51 \\
-\ 4.19 \\
\hline
\end{array}
$$

13.
$$
\begin{array}{r}
60.0782 \\
-\ 42.38 \\
\hline
\end{array}
$$

14.
$$
\begin{array}{r}
95.786 \\
-\ 88.999 \\
\hline
\end{array}
$$

15.
$$
\begin{array}{r}
4.9 \\
-\ 0.807 \\
\hline
\end{array}
$$

16.
$$
\begin{array}{r}
37.980 \\
-\ 37.421 \\
\hline
\end{array}
$$

17. $11.2 - 6.1356 =$

18. $29.006 - 8.0005 =$

19. $345.5842 - 0.095 =$

Copyright by The Goodheart-Willcox Co., Inc.

20. A pipe has a wall thickness of 0.129″. If the inside diameter is 3.716″, find the outside diameter.

21. A bar of cast iron 18.125″ long has been tapered from a diameter of 0.983″ to a diameter of 2.74″. What is the difference in diameter between the two ends?

22. Kendan Stamping produces a metal oil pan for a cost of $12.97. To make a profit of $4.98, at what price should it be sold?

23. Two pieces of metal measuring 65.283″ and 23.014″ long are welded together using a root opening of 0.031″. What length is the final piece?

Use this information to answer the two questions that follow. A salesperson estimated traveling expenses for May at $675.00. At the end of May, the receipts were as follows: gasoline: $120.59; meals: $298.97; room: $315.83; miscellaneous: $37.16.

24. What were the total traveling expenses?

25. By how much were the expenses greater or less than the estimated amount? Indicate a greater amount with a plus (+) and a lesser amount with a minus (−).

26. A steel pipe 17.074″ long has a 0.250″ cap welded to each end. The caps are then machined so that 0.0625″ of material are removed from each cap. What is the final length of the weldment?

Use the diagram below and this information to answer the questions that follow. Each rod in this weldment is 0.6534″ longer than the one before it. Rod A is 0.7596″ long.

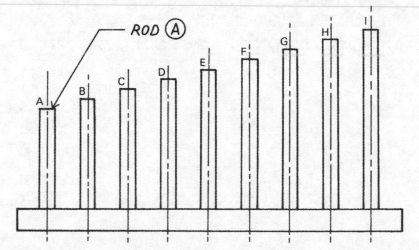

27. Calculate the length of Rod B.

28. Calculate the length of Rod C.

29. Calculate the length of Rod D.

Copyright by The Goodheart-Willcox Co., Inc.

Name _____ Date _____ Class _____

30. Calculate the length of Rod E.

31. Calculate the length of Rod F.

32. Calculate the length of Rod G.

33. Calculate the length of Rod H.

34. Calculate the length of Rod I.

35. Calculate the total length of rod used in this weldment.

36. Calculate the length of (A).

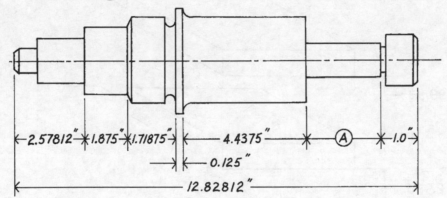

Use this information to answer the questions that follow. Review the diagram below. Grind 0.01562″ from one side of each block and 0.02813″ from the opposite side.

37. Calculate the finished thickness of block A.

38. Calculate the finished thickness of block B.

39. Calculate the finished thickness of block C.

40. Calculate the combined finished thickness of all three blocks.

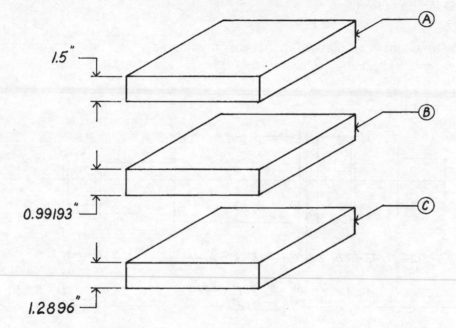

Copyright by The Goodheart-Willcox Co., Inc.

Name _____ Date _____ Class _____

41. Calculate the combined length of the six contact surfaces.

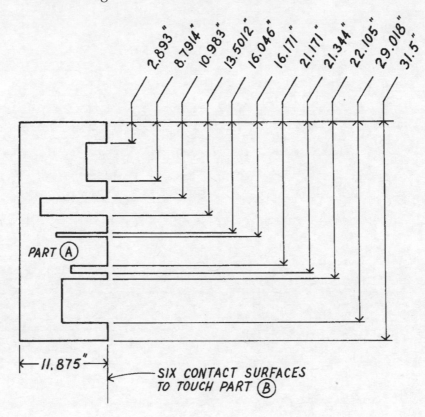

42. What are the overall dimensions of the block after 0.892″ is machined from each surface?

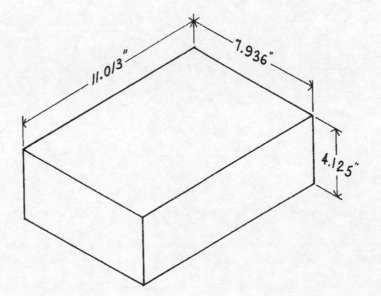

Copyright by The Goodheart-Willcox Co., Inc.

Work Space

Copyright by The Goodheart-Willcox Co., Inc.

Unit 13
Multiplication of Decimal Fractions

Introduction

Multiplying decimal numbers is almost exactly the same as multiplying whole numbers. Refer back to Unit 4, *Multiplication of Whole Numbers*, if you need to refresh yourself on the topic of multiplication.

Method Used to Multiply Decimal Fractions

When multiplying decimal numbers, the decimal points do not have to be lined up. Instead, you must line up the digits on the right side, as follows:

$$
\begin{array}{r}
14.124 \\
\times \quad .31 \\
\end{array}
$$

Now, multiply in the same way you would for whole numbers. You can ignore the decimal point until you complete the multiplication process. Remember to use an X or 0 as a marker to align the columns of the preliminary product of individual multiplier digits.

$$
\begin{array}{r}
14.124 \\
\times \quad .31 \\
\hline
14124 \\
42372X \\
\hline
437844 \\
\end{array}
$$

When you have arrived at an answer, you will have to figure out where the decimal should be placed. To do this, count the total number of figures to the right of the decimal point in the two numbers being multiplied. In this example, the total is five. Now, count off five figures from the right in the answer and place the decimal in front of (or, to the left of) the fifth figure. The answer here is 4.37844.

Occasions arise where you do not have enough digits in the answer to locate the decimal point. See the example below.

$$
\begin{array}{r}
.0234 \\
\times \quad .2 \\
\hline
0468 \\
\end{array}
$$

This problem contains a total of five figures to the right of the decimal, but the answer produces only four figures. Whenever this occurs, add as many zeros as needed to the left of the answer. The answer here is 0.00468.

A different situation is illustrated by the next example.

$$
\begin{array}{r}
.875 \\
\times \quad 2.4 \\
\hline
3500 \\
1750X \\
\hline
2.1000 \\
\end{array}
$$

When the answer ends in zeros on the right side of the decimal point, they may be eliminated. The answer here is 2.1.

Copyright by The Goodheart-Willcox Co., Inc.

This final example illustrates both adding zeros to the left and removing zeros from the right of the answer.

$$
\begin{array}{r}
.00684 \\
\times \quad .25 \\
\hline
03420 \\
01368X \\
\hline
017100
\end{array}
$$

The answer is 0.00171.

Multiplication of Decimals by 10, 100, 1,000, etc.

When a decimal number is multiplied by 10, 100, 1,000, 10,000, etc., the answer can be quickly calculated. Move the decimal to the <u>right</u> the same number of places as there are zeros in the multiplier. These examples will illustrate:

$$
\begin{array}{rcl}
1.8 \times 10 &=& 18.0 \\
1.8 \times 100 &=& 180.0 \\
1.8 \times 1,000 &=& 1,800.0 \\
.645 \times 100 &=& 64.5 \\
2,600,000 \times 10,000 &=& 26,000,000,000
\end{array}
$$

Copyright by The Goodheart-Willcox Co., Inc.

Name _____ Date _____ Class _____

Unit 13 Practice

Perform the following equations. Show all your work. Box your answers.

1. 0.89
    × 0.34

2. 0.409
    × 9.04

3. 0.572
    × 123

4. 7.008
    × 0.006

5. 10.0001
    × 100.1

6. 875
    × 0.25

7. $37 \times 0.18 =$

8. $18.125 \times 29 =$

9. $2{,}500 \times 0.4375 =$

10. $0.632 \times 22 =$

11. $0.52 \times 6.2 \times 9.3 =$

12. $0.66 \times 9.9 \times 12.0 =$

13. $0.02 \times 2.0 \times 0.002 =$

14. $0.5 \times 5 \times 0.05 =$

15. $569.59 \times 0.002 =$

16. $9.7825 \times 1.39 =$

17. $100{,}000 \times 1.95 =$

18. $0.0015 \times 100 =$

19. $2.423 \times 8.97 \times 3.0 =$

20. $10.01 \times 1{,}000 \times 1.0 =$

21. $11.1 \times 1.11 \times 0.111 =$

22. $22 \times 2.2 \times 0.22 =$

Copyright by The Goodheart-Willcox Co., Inc.

23. What is the weight of 2,906 aluminum castings if each casting weighs 23.47 pounds? Round your answer to the nearest ten pounds.

24. One cubic inch of steel weighs 0.2835 pounds. What is the weight of a block of steel containing 35.48 in³? Round your answer to the nearest hundredth.

25. A salesperson offered you the following deal on some shop supplies: $465.00 in cash or an $80.00 down payment with 12 monthly payments of $33.95 each. How much money would you save by paying cash?

26. What is the total cost to fill a storage tank with 5,385.5 gallons of diesel fuel at $4.87 per gallon? Express your answer in dollars and cents to the nearest cent.

27. A pipe support is welded in 3.75 minutes and uses 4.3125 ft³ of acetylene gas. How many cubic feet of gas are used to weld 12 of the supports?

28. An oxyacetylene cutting machine with 8 cutting torches uses 2.975 ft³ of oxygen per torch to produce 8 cover blanks. How many cubic feet of oxygen are required to produce 8,000 cover blanks?

29. A machining process at the Windsor Transmission Plant reduces a solid steel shaft from 103.301 pounds to 98.891 pounds. Every 60 minutes 351.75 shafts are produced. At the end of 53.6 hours, what is the total weight of steel removed from the shafts?

Use this information to answer the three questions that follow. Ten pieces of chain, each 6′ long, are required for a job. To clear out old inventory, the supply shop makes you a special offer: buy the remaining 64.5 feet of chain on the spindle at $3.74 per foot. Otherwise, you could buy only what you need at $3.91 per foot plus 60 cents ($0.60) per cut.

30. What is the total cost of the special offer?

31. What is the total cost of buying just what you need?

32. What is the cost saved by choosing the lower priced option?

33. Review the diagram below. Seven tube assemblies are required to complete a job. What is the total length of condenser tube required?

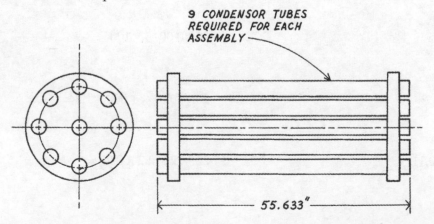

9 CONDENSOR TUBES
REQUIRED FOR EACH
ASSEMBLY

55.633″

Copyright by The Goodheart-Willcox Co., Inc.

Name _____ Date _____ Class _____

34. Review the diagram below. Calculate the total weight of the weldments in Job #8861.

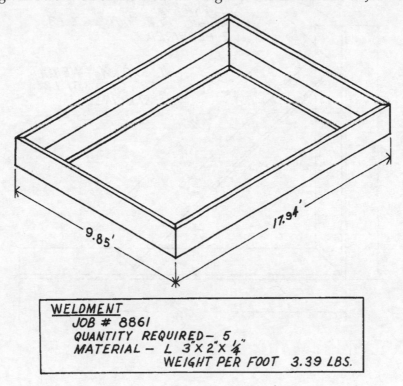

WELDMENT
  JOB # 8861
  QUANTITY REQUIRED— 5
  MATERIAL — L 3"X 2"X 1/4"
                WEIGHT PER FOOT 3.39 LBS.

35. Review the diagram below. Calculate the length of (A).

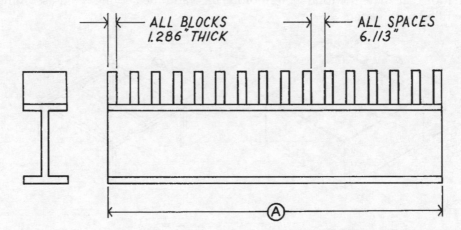

Copyright by The Goodheart-Willcox Co., Inc.

36. Review the diagram below. Twenty of these braces are required. What is the total weight?

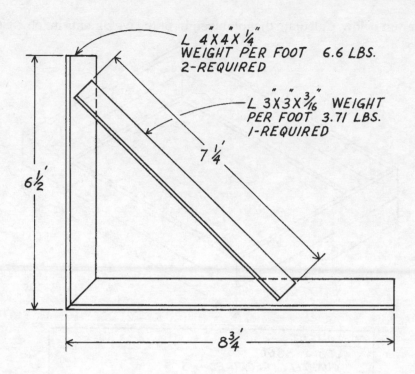

L 4"X4"X 1/4"
WEIGHT PER FOOT   6.6 LBS.
2-REQUIRED

L 3"X3"X 3/16" WEIGHT
PER FOOT 3.71 LBS.
1-REQUIRED

7 1/4'

6 1/2'

8 3/4'

37. Review the diagram below. The tank is made of 0.125″ thick steel. A 4′ wide sheet of steel weighs 20.412 pounds per foot. The base is made of 2″ square tubing with a wall thickness of .25″, which weighs 5.41 lb per foot. Water weighs 62.428 lb per cubic foot. The tank holds 26.56 ft³. Given this information, calculate the total weight of the base and the cooling tank, assuming that it is completely filled with water.

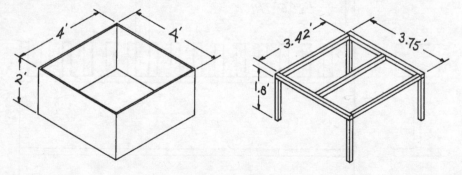

4'   4'

2'

3.42'   3.75'

1.8'

38. Review the following order and the following reference guide. Should a truck designed for a maximum load of 10,000 pounds be used to deliver the following order in one trip?

20 pieces of 1⁹⁄₁₆″ round bar at 12 feet

500 pieces of ¼″ round bar at 9 feet

70 pieces of 2″ square bar at 6 feet

5 pieces of 2¹³⁄₁₆″ round bar at 8 feet

Copyright by The Goodheart-Willcox Co., Inc.

Name _____ Date _____ Class _____

Square and Round Bars
Weight and Area

Size (in)	Weight (lb per foot) ■	Weight (lb per foot) ●	Area (in²) ▨	Area (in²) ◯	Size (in)	Weight (lb per foot) ■	Weight (lb per foot) ●	Area (in²) ▨	Area (in²) ◯
0					**3**	30.63	24.05	9.000	7.069
1/16	0.013	0.010	0.0039	0.0031	1/16	31.91	25.07	9.379	7.366
1/8	0.053	0.042	0.0156	0.0123	1/8	33.23	26.10	9.766	7.670
3/16	0.120	0.094	0.0352	0.0276	3/16	34.57	27.15	10.160	7.980
1/4	0.213	0.167	0.0625	0.0491	1/4	35.94	28.23	10.563	8.296
5/16	0.332	0.261	0.0977	0.0767	5/16	37.34	29.32	10.973	8.618
3/8	0.479	0.376	0.1406	0.1105	3/8	38.76	30.44	11.391	8.946
7/16	0.651	0.512	0.1914	0.1503	7/16	40.21	31.58	11.816	9.281
1/2	0.851	0.668	0.2500	0.1963	1/2	41.68	32.74	12.250	9.621
9/16	1.077	0.846	0.3164	0.2485	9/16	43.19	33.92	12.691	9.968
5/8	1.329	1.044	0.3906	0.3068	5/8	44.71	35.12	13.141	10.321
11/16	1.608	1.263	0.4727	0.3712	11/16	46.27	36.34	13.598	10.680
3/4	1.914	1.503	0.5625	0.4418	3/4	47.85	37.58	14.063	11.045
13/16	2.246	1.764	0.6602	0.5185	13/16	49.46	38.85	14.535	11.416
7/8	2.605	2.046	0.7656	0.6013	7/8	51.09	40.13	15.016	11.793
15/16	2.991	2.349	0.8789	0.6903	15/16	52.76	41.43	15.504	12.177
1	3.403	2.673	1.0000	0.7854	**4**	54.44	42.76	16.000	12.566
1/16	3.841	3.017	1.1289	0.8866	1/16	56.16	44.11	16.504	12.962
1/8	4.307	3.382	1.2656	0.9940	1/8	57.90	45.47	17.016	13.364
3/16	4.798	3.769	1.4102	1.1075	3/16	59.67	46.86	17.535	13.772
1/4	5.317	4.176	1.5625	1.2272	1/4	61.46	48.27	18.063	14.186
5/16	5.862	4.604	1.7227	1.3530	5/16	63.28	49.70	18.598	14.607
3/8	6.433	5.053	1.8906	1.4849	3/8	65.13	51.15	19.141	15.033
7/16	7.032	5.523	2.0664	1.6230	7/16	67.01	52.63	19.691	15.466
1/2	7.656	6.013	2.2500	1.7671	1/2	68.91	54.12	20.250	15.904
9/16	8.308	6.525	2.4414	1.9175	9/16	70.83	55.63	20.816	16.349
5/8	8.985	7.057	2.6406	2.0739	5/8	72.79	57.17	21.391	16.800
11/16	9.690	7.610	2.8477	2.2365	11/16	74.77	58.72	21.973	17.257
3/4	10.421	8.185	3.0625	2.4053	3/4	76.78	60.30	22.563	17.721
13/16	11.179	8.780	3.2852	2.5802	13/16	78.81	61.90	23.160	18.190
7/8	11.963	9.396	3.5156	2.7612	7/8	80.87	63.51	23.766	18.665
15/16	12.774	10.032	3.7539	2.9483	15/16	82.96	65.15	24.379	19.147
2	13.611	10.690	4.0000	3.1416	**5**	85.07	66.81	25.000	19.635
1/16	14.475	11.369	4.2539	3.3410	1/16	87.21	68.49	25.629	20.129
1/8	15.366	12.068	4.5156	3.5466	1/8	89.38	70.20	26.266	20.629
3/16	16.283	12.788	4.7852	3.7583	3/16	91.57	71.92	26.910	21.135
1/4	17.227	13.530	5.0625	3.9761	1/4	93.79	73.66	27.563	21.648
5/16	18.197	14.292	5.3477	4.20000	5/16	96.04	75.43	28.223	22.166
3/8	19.194	15.075	5.6406	4.4301	3/8	98.31	77.21	28.891	22.691
7/16	20.217	15.879	5.9414	4.6664	7/16	100.61	71.92	29.566	23.221
1/2	21.267	16.703	6.2500	4.9087	1/2	102.93	80.84	30.250	23.758
9/16	22.344	17.549	6.5664	5.1572	9/16	105.29	82.69	30.941	24.301
5/8	23.447	18.415	6.8906	5.4119	5/8	107.67	84.56	31.641	24.850
11/16	24.577	19.303	7.2227	5.6727	11/16	110.07	86.45	32.348	25.406
3/4	25.734	20.211	7.5625	5.9396	3/4	112.50	88.36	33.063	25.967
13/16	26.917	21.140	7.9102	6.2126	13/16	114.96	90.29	33.785	26.535
7/8	28.126	22.090	8.2656	6.4918	7/8	117.45	92.24	34.516	27.109
15/16	29.362	23.061	8.6289	6.7771	15/16	119.96	94.22	35.254	27.688
3	30.625	24.053	9.0000	7.0686	**6**	122.50	96.21	36.000	28.274

American Institute of Steel Construction

Copyright by The Goodheart-Willcox Co., Inc.

39. Review the parts list below. What is the total weight of Job #8853?

Parts List
Job—8853
Date—August 29

Quantity	Part Description	Weight
37	3/4″ STD pipe at 6.25′	1.13 lb per foot
148	1″ round bar at 0.5685′	2.67 lb per foot
4	M shape at 10.5′	18.9 lb per foot
7	L 5″ × 3″ × 1/4″ at 7.0625′	6.6 lb per foot
8	3″ STD channel at 8.125′	4.1 lb per foot

Copyright by The Goodheart-Willcox Co., Inc.

Unit 14
Division of Decimal Fractions

Introduction

Dividing decimal numbers is almost exactly the same as dividing whole numbers. Refer back to Unit 5, *Division of Whole Numbers*, if you need to review the topic of division.

Method Used to Divide Decimal Fractions

When dividing decimal numbers, arrange dividend and divisor as you would for whole number division. However, be mindful that your first concern is the decimal point. Here is what you must do. If there is a decimal point in the divisor, simply move it to the right of the far right digit. You want to change the divisor to a whole number. Then move the decimal point in the dividend the same number of places to the right. If necessary, add zeros to the right of the decimal. Here is an example.

Divide 186.942 by 17.34 to the nearest thousandth. Begin with the division equation setup.

$$17.34\overline{)186.942}$$

Move the decimal in the divisor (17.34) to the far right and then move the decimal in the dividend (186.942) the same number of places to the right.

$$1734.\overline{)18694.2}$$

This decimal point shifting changes the divisor to a whole number and greatly simplifies the task of division. Next, place a decimal directly above the decimal in the dividend. This is done so the decimal in the answer is properly located.

$$1734.\overline{)18694.2}$$

Now, proceed with the division. Remember to calculate to the ten thousandth so the answer can be rounded to the nearest thousandth.

```
            10.7809
  1734. )18694.2000
       - 1734
         1354
       -    0
         1354 2
       - 1213 8
          140 40
        - 138 72
           1 680
         -     0
           1 6800
         - 1 5606
             1194
```

The quotient 10.7809 is then rounded to the nearest thousandth. The 9 in the ten thousandth place rounds the 0 in the thousandth place up to 1. The answer is 10.781.

Copyright by The Goodheart-Willcox Co., Inc.

As shown in the example, the dividend may not have enough zeros to accommodate the relocated decimal. If this occurs, add enough zeros so you can locate the decimal properly. See the example below.

$$9.345 \overline{\smash{\big)}27.25}$$

The decimal is relocated like this:

$$9345. \overline{\smash{\big)}27250.}$$

Such arrangements may occur commonly but are easily remedied by careful column alignment and attention to detail.

Division of Decimals by 10, 100, 1,000, etc.

When a decimal number is divided by 10, 100, 1,000, 10,000, etc., the answer can be quickly calculated. Move the decimal to the <u>left</u> the same number of places as there are zeros in the divisor. These examples will illustrate:

$$185 \div 10 = 18.5$$
$$185 \div 100 = 1.85$$
$$185 \div 1{,}000 = .185$$
$$968 \div 10{,}000 = .0968$$
$$7{,}500{,}000 \div 1{,}000{,}000 = 7.5$$

Copyright by The Goodheart-Willcox Co., Inc.

Name _____   Date _____   Class _____

Unit 14 Practice

Divide the following equations. Show all your work. Box your answers.

1. $875 \div 0.25 =$

2. $1.17 \div 0.003 =$

3. $3.94 \div 20 =$

4. $10.001 \div 0.01 =$

5. $0.072 \div 0.009 =$

6. $8.6752 \div 1{,}000 =$

Divide the following equations to the nearest hundredth. Show all your work. Box your answers.

7. $\dfrac{0.1217}{1.72} =$

8. $\dfrac{7.85}{2.16} =$

9. $\dfrac{100.449}{100} =$

10. $\dfrac{96{,}603}{100{,}000} =$

11. $\dfrac{6.3}{0.046} =$

12. $\dfrac{72}{0.71} =$

Divide the following equations to the nearest thousandth. Show all your work. Box your answers.

13. $16.5 \div 5.5 =$

14. $48 \div 10.1 =$

15. $17.1298 \div 1{,}000 =$

16. $0.96 \div 6.9 =$

17. $6.9 \div 9.6 =$

18. $2{,}009.15 \div 100 =$

19. A gear pump can move 113.4 gallons of water per hour. How many hours would it take to empty a tank containing 1,822 gallons? Express your answer to the nearest hour.

20. One carton containing 9 one-gallon cans of cold galvanizing compound cost $137.25. How many one-gallon cans are you able to buy for $4,422.50?

21. A large container of stainless steel ball bearings weighs 212.55 pounds. Each ball bearing weighs 0.2834 pounds. How many ball bearings are in the container?

22. A surface grinder removes 0.013″ of steel with each pass. How many passes are required to reduce a piece of steel from 2.095″ to 1.718″?

Copyright by The Goodheart-Willcox Co., Inc.

Use this information to answer the two questions that follow. A special assignment includes producing precision cut blocks of steel measuring 0.869″ thick. These blocks are going to be placed into an opening that is 3.652″.

23. How many precision-cut blocks will fit in the opening?

24. How much space, if any, will remain in the opening after the blocks have been inserted?

25. A company offered the following deal on a standard item they produced. The first 150 items a customer ordered would cost $5,227.50. Each item after that would cost the customer $30.50. If the company received a $19,654.00 order, how many items were in the order?

26. Refer to the diagram below. Divide this channel into 11 equal pieces. What is the length of each piece?

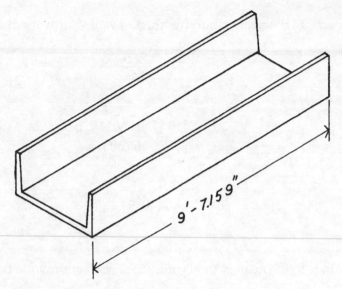

9′ - 7.159″

27. How many 1.857″ squares can be cut from this sheet of steel?

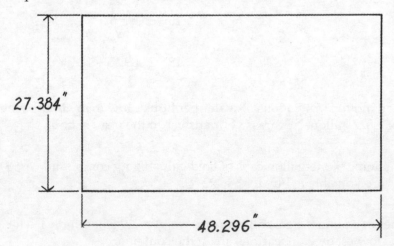

27.384″

48.296″

Copyright by The Goodheart-Willcox Co., Inc.

Name _____ Date _____ Class _____

28. This square bar weighs 156.686 pounds. What is the weight of a 1" piece?

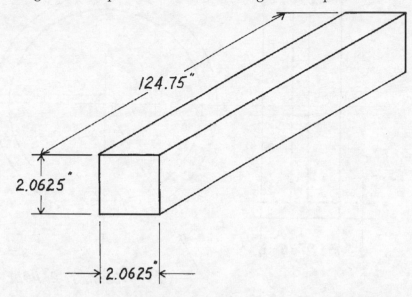

29. The clearance between the collar and drum shown below is 0.0573". Using a thermal coating process, the clearance is to be reduced to 0.0087". Each coating application deposits 0.0009" of material. How many applications are required?

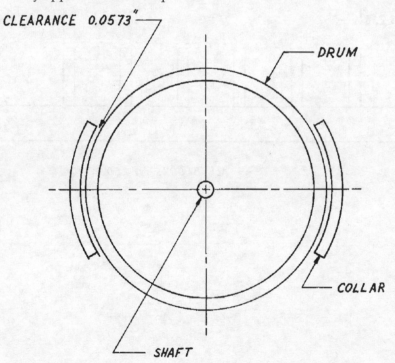

Copyright by The Goodheart-Willcox Co., Inc.

30. How many of the following drums would there be in a stack 73.0232″ high?

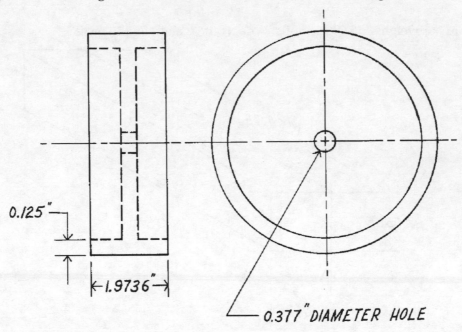

0.125″

1.9736″

0.377″ DIAMETER HOLE

31. What is the distance between the evenly spaced vertical blocks?

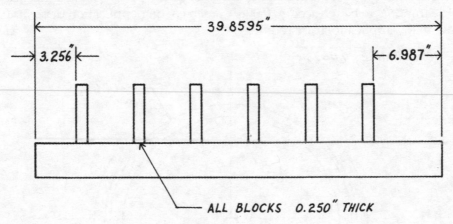

39.8595″

3.256″

6.987″

ALL BLOCKS 0.250″ THICK

Copyright by The Goodheart-Willcox Co., Inc.

Section 4

Measurement

Section Objectives

After studying this section, you will be able to:

- Define linear measure, angular measure, and circular measure
- Show how to convert linear measurements between units
- Explain and give examples of various forms of tolerance
- Show how to perform angular measurement
- Perform four basic operations for angular calculations
- Define basic four-sided shapes
- Calculate perimeter and area for four-sided shapes
- Convert square units of measure
- Define basic triangular shapes
- Calculate perimeter and area for triangular shapes
- Define various parts of circular shapes
- Calculate circumference and area of circular shapes

Copyright by The Goodheart-Willcox Co., Inc.

Unit 15
Linear Measure

Key Terms

angular measure   linear measure   SI (metric) system   US Customary system

circular measure   millimeter (mm)   tolerance (±)

Introduction

A welder's job includes taking and reading measurements. Generally, as a welder, you will be dealing with the three types of measurements listed below:

- **Linear measure** refers to measuring the straight line distance between two points.
- **Angular measure** refers to measuring the angle formed by two intersecting lines.
- **Circular measure** refers to measuring curved lines.

Two linear measuring systems are presently in use by welders, the US Customary system and the SI (metric) system. You should become very familiar with both systems.

US Customary System

The **US Customary system** is the most common measuring system with which we are all familiar. The basic unit of length is the inch. Twelve inches make up one foot and 36 inches form one yard. In welding, you will be concerned only with feet and inches. For example, a standard notation is 6'-11". This notation lists feet before inches and connects them with a hyphen (-).

SI Metric System

You are taking your training just at the time the welding industry, along with the rest of industrial America, is changing from the US Customary system to the **SI (metric) system**. A full explanation of metrics as it relates to your work is provided in Unit 24, *The Metric System*. A short explanation of metric linear measurement follows.

The Millimeter

The smallest unit of length you will encounter in the shop or in the field is the **millimeter**. In fact, it is probably the only metric unit you will use because almost all metric blueprints are dimensioned only in millimeters. It is the basic unit in metric dimensioning, just as the inch is the basic unit in the US Customary system. The notation used for millimeters is **mm**. As you can see in the illustration below, it is a very small unit of measurement.

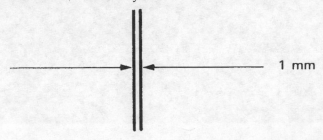

1 mm

Copyright by The Goodheart-Willcox Co., Inc.

Here are two ways to visualize a millimeter.

- A millimeter is approximately the thickness of a line made with a dull pencil.
- A dime is about one millimeter thick.

Since millimeters are very small, they accumulate very quickly. For example, there are 25.4 millimeters in one inch. If you are 5′-10″ tall, then you are 1778 mm tall. A 10′ bar is 3048 mm long. Eventually, through practice, you will develop a "feel" for metric lengths. If you do not own a metric scale, you should acquire one now. Note that in the metric system, a comma is not placed between sets of three digits.

Converting Linear Measurements

Your shop work will require you to change back and forth between inches, feet, and millimeters. Often, these conversions will result in lengthy decimals, so it is common practice to round off conversions. Also, you will sometimes have to convert decimal dimensions to fractional dimensions normally found on your scale (eighths, sixteenths, etc.).

Converting Feet to Inches

To convert from feet to inches, multiply the number of feet by twelve.

$$7' = 7 \times 12 = 84''$$

$$27.35' = 27.35 \times 12 = 328.2''$$

$$4\frac{19}{64}' = \frac{275}{64} \times 12 = \frac{3,300}{64} = 51\frac{36}{64} = 51\frac{9}{16}''$$

Converting Inches to Feet

To convert from inches to feet, divide the number of inches by twelve. The answer is in its most useful form when expressed in feet and inches rather than only in feet. Note that the remainder of each division problem is the number of inches to be written after the number of feet in the notation.

$$109'' = 109 \div 12 = 9'\text{-}1''$$

$$219\frac{1}{4}'' = 219\frac{1}{4} \div 12 = 18'\text{-}3\frac{1}{4}''$$

$$28.7'' = 28.7 \div 12 = 2'\text{-}4.7''$$

As shown in the last two examples, the partial inch values did not need to be included in the calculation. This would have complicated the equation. These fractional inch values were easily brought over from the original value. Only whole inches need to be used in these conversion calculations, because only whole inches can be used to make up a foot measurement.

Converting Inches to Millimeters

To convert from inches to millimeters, multiply the number of inches by 25.4. A good rule for rounding the answer is to express it one <u>less</u> decimal place than the question.

$$16'' = 16 \times 25.4 = 406.4 = 406 \text{ mm}$$

$$93.375'' = 93.375 \times 25.4 = 2371.725 = 2371.73 \text{ mm}$$

$$\frac{1}{16} = \frac{1}{16} \times 25.4 = 1.5875 = 1.6 \text{ mm}$$

Copyright by The Goodheart-Willcox Co., Inc.

Converting Millimeters to Inches

To convert from millimeters to inches, divide the number of millimeters by 25.4. A good rule for rounding the answer is to express it one <u>more</u> decimal place than the question.

$$1789 \text{ mm} = 1789 \div 25.4 = 70.4''$$

$$9.32 \text{ mm} = 9.32 \div 25.4 = .3669 = .367''$$

$$360.5 \text{ mm} = 360.5 \div 25.4 = 14.19''$$

Summary of Conversion Factors		
From	**To**	**Method**
Feet	Inches	Multiply by 12
Inches	Feet	Divide by 12
Inches	Millimeters	Multiply by 25.4
Millimeters	Inches	Divide by 25.4

Converting Decimals to Fractions

In Unit 11, *Introduction to Decimal Fractions*, you learned to convert decimals to fractions. A new situation arises when doing such a conversion for the purpose of linear measure. Here, you will have to convert the decimal dimension to one of the fractional dimensions normally found on your steel rule, such as eighths, sixteenths, thirty-seconds, or sixty-fourths. A typical example would be to convert 10.21" to a fractional dimension to the nearest sixty-fourth.

1. Dealing with the decimal part only, 0.21 converts to $\frac{21}{100}$.

2. Since $\frac{21}{100}$ will not reduce exactly to a fraction with a denominator of 64, it will have to be reduced to the nearest sixty-fourth.

3. Write an equation as shown:

$$\frac{21}{100} = \frac{x}{64}$$

Since you want the final fraction to be in sixty-fourths, you know the denominator will be 64. However, you do not yet know what the numerator will be, so it is identified for now as "x."

4. The method for finding x can be expressed in one phrase: <u>Cross multiply and divide</u>.

   1. $\dfrac{21}{100} \diagdown \dfrac{x}{64}$

   2. $21 \times 64 = 1344$

   3. $1344 \div 100 = 13\dfrac{11}{25}$

Remember, you are trying to find the number that will replace the x in the fraction $\frac{x}{64}$. That number as calculated so far would be $13\frac{11}{25}$. But, instead of using $13\frac{11}{25}$, round to the nearest whole number, which is 13, since 11 is closer to 0 than to 25. Thirteen, then, is the numerator.

5. The answer is $10\frac{13}{64}''$ (to the nearest sixty-fourth).

Copyright by The Goodheart-Willcox Co., Inc.

Tolerance

Taking accurate measurements is obviously important to any tradesperson. But there is a limit to how accurately you can measure. This limit is determined by two main factors: the accuracy of the measuring tool and the accuracy of the surface of the material being measured. Because of this, dimensions on blueprints normally indicate a certain acceptable range that you are allowed to work within. This range is called the **tolerance**, and it is usually indicated as shown in the following examples.

Tolerance +⅛″

This indicates an object may exceed the given dimension by ⅛″. Given a tolerance of +⅛″, an object dimensioned 48⅜″ long would have a maximum acceptable dimension of 48½″ and a minimum of 48⅜″.

Tolerance −0.0625″

This indicates an object may be 0.0625″ less than the given dimension. Given a tolerance of −0.0625″, an object dimensioned 99.6″ long would have a maximum acceptable dimension of 99.6″ and a minimum of 99.5375″.

Tolerance ±3 mm

This indicates an object may be 3 mm greater or less than the given dimension. In this case, you have a range of 6 mm to work within. Given a tolerance of ±3 mm, an object dimensioned 211 mm long would have a maximum acceptable dimension of 214 mm and a minimum of 208 mm.

Copyright by The Goodheart-Willcox Co., Inc.

Work Space

Copyright by The Goodheart-Willcox Co., Inc.

Name _____ Date _____ Class _____

Unit 15 Practice

Measure the following lines as accurately as possible, in millimeters and inches. Box your answers.

1. ━━━━━━━━━━━━━━━━━━━━━━━

2. ━━

3. ━━━━━━━━━━━━━━━

4. ━━━━━━━━━━━━━━━━━━━━━━━━━━

5. ━━━━━

6. ━━━━━━━━━━━━

Measure the following as accurately as possible in millimeters and inches.

7. Your height: _____, _____

8. The height of a coffee cup: _____, _____

9. The length of a welding rod: _____, _____

10. The height of a workbench or desk: _____, _____

11. The span of your hand from smallest finger to thumb: _____, _____

12. The width of a doorway: _____, _____

Convert the following values from feet to inches. Show all your work. Box your answers.

13. 35¼′   14. 123′

15. 45.375′   16. 17⁵/₂₃′

17. 503.5′   18. 4⅛′

19. 5,280′   20. 65.1′

21. ¹⁷/₆₄′   22. 29.47′

23. 12¹/₁₂′   24. ¹⁵/₃₂′

Copyright by The Goodheart-Willcox Co., Inc.

Convert the following values of inches to feet (and inches, if necessary). Show all your work. Box your answers.

25. 1½″

26. 6,000″

27. 95.75″

28. ⅔″

29. 215¹⁄₃₂″

30. 12″

31. 787.96″

32. ¹⁷⁄₆₄″

33. 83,917″

34. 38¼″

35. 4⅛″

36. 0.29″

Convert the following inch values to millimeters. Round to the nearest tenth of a millimeter. Show all your work. Box your answers.

37. 12″

38. 36″

39. 10″

40. 1″

41. ¹⁄₆₄″

42. 69″

43. ⅝″

44. 1,000″

45. 63³⁄₁₆″

46. 59″

47. 8¼″

48. 480″

Convert the following millimeters to the decimal equivalent in inches. Round to the nearest thousandth of an inch. Show all your work. Box your answers.

49. 1000 mm

50. 25.4 mm

51. 1.0 mm

52. 279.4 mm

53. 65.5 mm

54. 12 mm

55. 11850 mm

56. 35600 mm

57. 195 mm

58. 862.3 mm

59. 5280 mm

60. 100 mm

Copyright by The Goodheart-Willcox Co., Inc.

Name _____ Date _____ Class _____

Convert the following decimal inches to fractional dimensions. Calculate to the nearest eighth of an inch. Show all your work. Box your answers.

61. 15.8″

62. 0.982″

63. 48.012″

64. 80.125″

65. 19.375″

66. 0.9126″

Convert the values below to the nearest sixteenth of an inch. Show all your work. Box your answers.

67. 19.85″

68. 32.032″

69. 0.997″

70. 855.5″

71. 21.0625″

72. 1,010.7″

Convert to the nearest thirty-second of an inch. Show all your work. Box your answers.

73. 11.65″

74. 58.022″

75. 730.18″

76. 0.28125″

77. 0.979″

78. 3,602.79″

Convert to the nearest sixty-fourth of an inch. Show all your work. Box your answers.

79. 68.033″

80. 0.908″

81. 12.55″

82. 0.578125″

83. 632.41″

84. 7,802.77″

Copyright by The Goodheart-Willcox Co., Inc.

Perform calculations and fill in the blanks in the table below. Determine the maximum and minimum dimensions allowed in the following lengths.

	Dimension	Tolerance	Maximum	Minimum
85.	249 mm	– 2 mm		
86.	17¹⁄₃₂″	+ ¹⁄₁₆″		
87.	1.340″	± 0.005″		
88.	8′ 11¾″	+ ½″		
89.	14.5 mm	± 0.5 mm		
90.	18⁹⁄₁₆″	+ ¹⁄₃₂″		
91.	19.9 mm	± 0.2 mm		
92.	8′ 0″	– ¹⁄₁₆″		

Review the diagram below. Convert its dimensions to fractional dimensions (to the nearest one-hundredth of an inch). Then, using a tolerance of ±3/100″, calculate the maximum and minimum allowable dimensions. Record your answers in the table below.

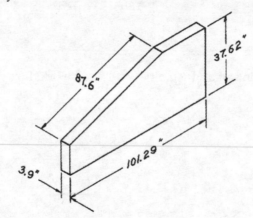

	Decimal Dimension	Fractional Dimension	Tolerance	Maximum	Minimum
93.	3.9″		± 3/100″		
94.	37.62″		± 3/100″		
95.	87.6″		± 3/100″		
96.	101.29″		± 3/100″		

Copyright by The Goodheart-Willcox Co., Inc.

Name _____ Date _____ Class _____

Review the diagram below. The tolerance for the hole in this weldment is +1/8". All other dimensions have a tolerance of ±1/16". Calculate the maximum and minimum allowable dimensions and record your answers in the table below.

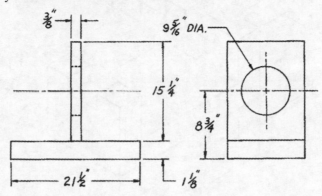

	Fractional Dimension	Tolerance	Maximum	Minimum
97.	⅜"	± ¹⁄₁₆"		
98.	1⅛"	± ¹⁄₁₆"		
99.	8¾"	± ¹⁄₁₆"		
100.	9⁵⁄₁₆"	+ ⅛"		
101.	15¼"	± ¹⁄₁₆"		
102.	21½"	± ¹⁄₁₆"		

Copyright by The Goodheart-Willcox Co., Inc.

Refer to the diagram below and perform calculations to fill in the blanks in the table that follows.

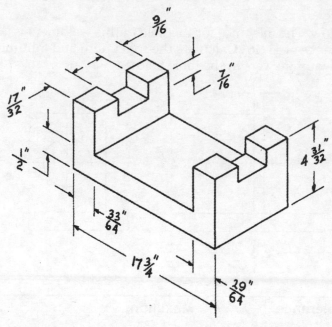

Prepare a list of decimal equivalents of all the dimensions from the diagram above. Round each decimal equivalent to three decimal places.

Prepare a list of the metric equivalents in millimeters of all the dimensions from the diagram above. Round each metric decimal equivalent to two decimal places.

	Fractional Dimensions	Decimal Equivalents	Metric Decimal Equivalents
103.	$^7/_{16}''$		
104.	$^{29}/_{64}''$		
105.	$^1/_2''$		
106.	$^{33}/_{64}''$		
107.	$^{17}/_{32}''$		
108.	$^9/_{16}''$		
109.	$4^{31}/_{32}''$		
110.	$17^3/_4''$		

Copyright by The Goodheart-Willcox Co., Inc.

Name _____ Date _____ Class _____

Review the diagram below. Using a tolerance of –.5 mm, calculate the maximum and minimum allowable dimensions. Record your answers in the table below.

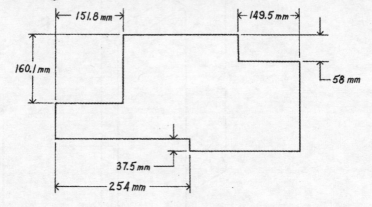

	Decimal Dimensions	Maximum	Minimum
111.	37.5 mm		
112.	58.0 mm		
113.	149.5 mm		
114.	151.8 mm		
115.	160.1 mm		
116.	254.0 mm		

Copyright by The Goodheart-Willcox Co., Inc.

Review the diagram below. Specified lengths are designated with a letter. Using a metric scale, measure each of these lengths to the nearest millimeter. Record these values in the table.

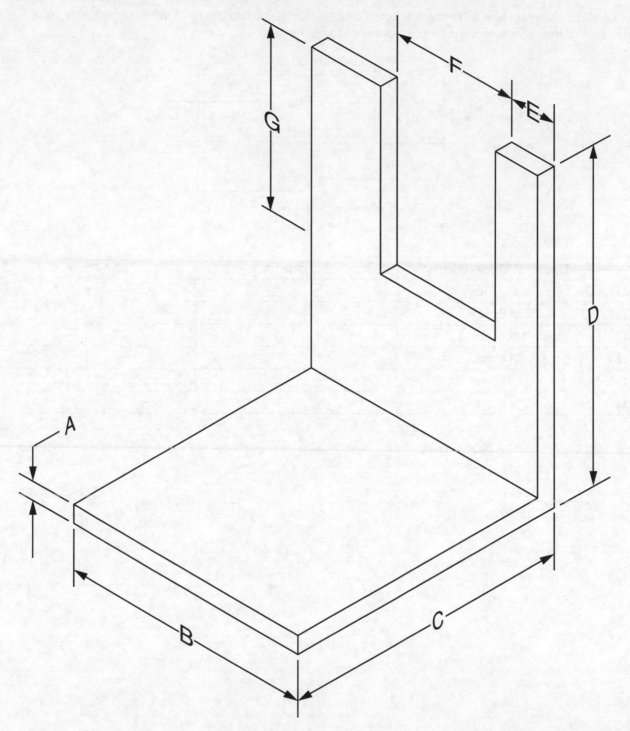

Copyright by The Goodheart-Willcox Co., Inc.

Name _____ Date _____ Class _____

	Letter Designations	Metric Decimal Measurements	Customary Decimal Calculations	Customary Fractional Calculations
117.	A			
118.	B			
119.	C			
120.	D			
121.	E			
122.	F			
123.	G			

Calculate the equivalent dimensions in decimal inches to the nearest thousandth of an inch. Show your work. Record these calculated values in the table.

Calculate the equivalent dimensions in fractional inches to the nearest sixty-fourth of an inch. Record these calculated values in the table.

Copyright by The Goodheart-Willcox Co., Inc.

Work Space

Copyright by The Goodheart-Willcox Co., Inc.

Unit 16
Angular Measure

Key Terms

angle   minute   second
degree   protractor

Introduction

An **angle** can be defined as the opening between two intersecting lines.

Welders are often called upon to work with angles. Therefore, you should be able to take measurements of angles and to perform math operations involving angles.

Angular Measurement

The units of measure used for angles are **degrees**, **minutes**, and **seconds**. The largest unit is the degree. Mathematicians have divided the circle into 360 parts, calling each part a degree. Therefore, 1 degree is ⅟₃₆₀ of a circle. The notation for degrees is °. To provide a more accurate measurement, each degree has been divided into 60 equal parts, called minutes, and noted as ′. To allow for even more accurate measurement, each minute has been subdivided into 60 equal parts called seconds, and noted as ″. (Unfortunately, the notation for minutes and seconds is the same as that used for feet and inches in linear measure.)

Angles can be measured using an instrument called a **protractor**, as shown below.

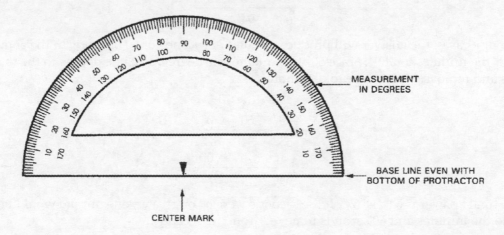

Copyright by The Goodheart-Willcox Co., Inc.

To use a protractor, align the base of the protractor along one line of the angle. Position the protractor's center marker at the intersection of the lines making the angle. Then, follow the line of the angle that does not run along the base of the protractor. This line will cross under the curved part of the protractor, showing the number value of the angle. In the following example, the protractor reads the angle as 55°.

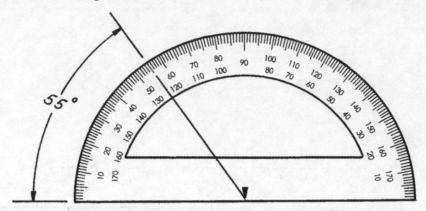

Since the degree of accuracy in measuring angles is limited by the tools being used and the material being measured, your shop work is normally allowed a tolerance. As with linear measure, the tolerance may be +, –, or ±. A typical example on a blueprint may call for an angle of 53° ±2°. In this case, the angle could vary from 55° to 51° and still be acceptable.

Angular Calculation

The basic math operations of addition, subtraction, multiplication, and division can be performed with angles.

Addition of Angles

When adding angles, align the units of measure (degrees, minutes, seconds) in columns as shown and add each unit, starting from the far right.

$$
\begin{array}{rrr}
45° & 39' & 17'' \\
+ \, 18° & 55' & 40'' \\
\hline
63° & 94' & 57'' \\
\end{array}
$$

In many cases, the answer will produce minutes or seconds of 60 or more. In the above example, the minutes total 94. Since 60' are equal to one degree, add one degree to the total degrees and remove 60' from the total minutes.

$$
\begin{array}{rrr}
63° & 94' & 57'' \\
- \, 0° & 60' & 0'' \\
\hline
63° & 34' & 57'' \\
+ \, 1° & 0' & 0'' \\
\hline
64° & 34' & 57'' \\
\end{array}
$$

The final answer is 64° 34′ 57″. If the seconds were 60 or greater, one minute would be added to the minutes and 60 seconds removed from the seconds.

Copyright by The Goodheart-Willcox Co., Inc.

Subtraction of Angles

As in addition of angle, align the units in columns for subtraction.

$$
\begin{array}{rrr}
11° & 52' & 10'' \\
-7° & 29' & 37'' \\
\hline
\end{array}
$$

In this example, 37″ cannot be subtracted from 10″. Borrow 1′ from the 52′, add 60″ to the 10″.

$$
\begin{array}{rrr}
11° & 52' & 10'' \\
-0° & 1' & 0'' \\
\hline
11° & 51' & 10'' \\
+0° & 0' & 60'' \\
\hline
11° & 51' & 70'' \\
\end{array}
$$

Now subtraction is possible without hindrance.

$$
\begin{array}{rrr}
11° & 51' & 70'' \\
-7° & 29' & 37'' \\
\hline
4° & 22' & 33'' \\
\end{array}
$$

When subtracting angles, be mindful of how many units will be borrowed between columns.

Multiplication of Angles

When multiplying angles, treat each column of units separately. Multiply each column of units by the multiplier and keep all the digits of each unit together.

$$
\begin{array}{rrr}
30° & 47' & 22'' \\
\times & & 4 \\
\hline
120° & 188' & 88'' \\
\end{array}
$$

After multiplication is finished, then reduce each column of units that has exceeded its high value limit. Begin this process from the right side of the angle number. In this example, subtract 60″ from the 88″ and add 1′ to the 188′ to arrive at 189′.

$$
\begin{array}{rrr}
120° & 188' & 88'' \\
-0° & 0' & 60'' \\
\hline
120° & 188' & 28'' \\
+0° & 1' & 0'' \\
\hline
120° & 189' & 28'' \\
\end{array}
$$

Notice the high value of the minute column. It only takes 60′ to equal 1°. Dividing 189 by 60 results in 3 with a remainder of 9. Since 3 is the number of times 60 fits into 189, the 3 is added to the degree column. The remainder of 9 is left over as the new minute value.

$$
\begin{array}{rrr}
120° & 189' & 28'' \\
-00° & 180' & 00'' \\
\hline
120° & 9' & 28'' \\
+03° & 0' & 0'' \\
\hline
123° & 9' & 28'' \\
\end{array}
$$

Copyright by The Goodheart-Willcox Co., Inc.

Division of Angles

While multiplication of angles begins with the smallest unit (seconds), division of angles begins with the largest unit (degrees). Each set of units is divided separately. After division of a column of units is complete, any remainders are converted to the next lower valued column's unit and added to the existing values. Then division of that column may begin.

$$
\begin{array}{r}
40° \qquad 34' \qquad 44'' \\
3\,\overline{|\,121° \qquad 44' \qquad 14''} \\
\underline{-12\!\downarrow} \\
01 \;\longrightarrow\; +60 \\
\overline{104} \\
\underline{-9\!\downarrow} \\
14 \\
\underline{-12} \\
2 \;\longrightarrow\; +120 \\
\overline{134} \\
\underline{-12\!\downarrow} \\
14 \\
\underline{-12} \\
2\ \text{r}
\end{array}
$$

When a remainder occurs, round it to the nearest second. To do so, use the remainder and divisor to form a fraction $\tfrac{2}{3}$. Dividing 2 by 3 approximately equals 0.67. Since 0.67 is closer to 1 than 0, the second's column number is rounded up. The answer is 40° 34′ 45″ rounded to the nearest second.

Copyright by The Goodheart-Willcox Co., Inc.

Name _____ Date _____ Class _____

Unit 16 Practice

Using a protractor, measure the following angles as accurately as possible. List your answers below.

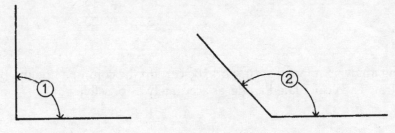

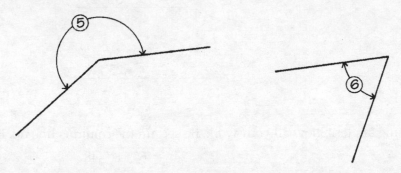

1. _____   2. _____

3. _____   4. _____

5. _____   6. _____

Add the following angles. Show all your work. Be certain the columns line up. Box your answers. Using a protractor, draw each calculated angle, as accurately as possible.

7.   7° 20′
   + 3° 18′

8.   92° 36′ 10″
   +   13′ 35″

9.   45° 45′ 9″
   + 29° 35′ 51″

10. 152° 32′ 54″
   + 27° 27′ 6″

Copyright by The Goodheart-Willcox Co., Inc.

11. 109° 51′ 44″
   + 90° 47′ 39″
   ‾‾‾‾‾‾‾‾‾‾‾‾‾‾

12.   91° 55′ 12″
   + 76° 2′ 14″
   ‾‾‾‾‾‾‾‾‾‾‾‾‾‾

13.   55° 13′ 16″
   + 11° 37′ 39″
   ‾‾‾‾‾‾‾‾‾‾‾‾‾‾

14.   19° 50′ 1″
        4° 6′ 19″
   + 32° 9′ 20″
   ‾‾‾‾‾‾‾‾‾‾‾‾‾‾

Subtract the following angles. Show all your work. Be certain the columns line up. Box your answers. Using a protractor, draw each calculated angle, as accurately as possible.

15. 119° 17′ 43″
    − 91° 15′ 23″
   ‾‾‾‾‾‾‾‾‾‾‾‾‾‾

16.   10° 40′
    − 7° 37′
   ‾‾‾‾‾‾‾‾‾‾‾‾‾‾

17. 135° 54′ 11″
    −101° 55′ 9″
   ‾‾‾‾‾‾‾‾‾‾‾‾‾‾

18. 204° 6′ 32″
    − 69° 3′ 42″
   ‾‾‾‾‾‾‾‾‾‾‾‾‾‾

19. 652° 44′ 0″
    −359° 0′ 18″
   ‾‾‾‾‾‾‾‾‾‾‾‾‾‾

20.   80° 22′ 31″
    − 79° 31′ 59″
   ‾‾‾‾‾‾‾‾‾‾‾‾‾‾

21.   45° 0′ 21″
    − 33° 7′ 53″
   ‾‾‾‾‾‾‾‾‾‾‾‾‾‾

22.   86° 36′ 12″
    −   23′ 14″
   ‾‾‾‾‾‾‾‾‾‾‾‾‾‾

Multiply the following angles. Show all your work. Be certain the columns line up. Box your answers.

23.   35° 24′ 22″
    ×   8
   ‾‾‾‾‾‾‾‾‾‾‾‾‾‾

24.   46° 11′ 48″
    ×   9
   ‾‾‾‾‾‾‾‾‾‾‾‾‾‾

25. 129° 58′ 36″
    ×   7
   ‾‾‾‾‾‾‾‾‾‾‾‾‾‾

26.   90° 43′
    ×   5
   ‾‾‾‾‾‾‾‾‾‾‾‾‾‾

27. 206° 32′ 47″
    ×   6
   ‾‾‾‾‾‾‾‾‾‾‾‾‾‾

28.   18° 14′ 3″
    ×   12
   ‾‾‾‾‾‾‾‾‾‾‾‾‾‾

Divide the following angles. Show all your work. Be certain the columns line up. Box your answers. Using a protractor, draw each calculated angle, as accurately as possible.

29. 3)‾393° 42′ 27″

30. 2)‾181° 13′ 50″

31. 6)‾180°

32. 8)‾222°

Copyright by The Goodheart-Willcox Co., Inc.

Name _____ Date _____ Class _____

33. $7 \overline{)360°}$

34. $8 \overline{)360°}$

35. $11 \overline{)93° \quad 7' \quad 45''}$

36. $27 \overline{)19° \quad 14' \quad 57''}$

Review the information in the table below. Calculate the maximum and minimum angles according to the given tolerance and record these values in the table below.

	Angle	Tolerance	Maximum	Minimum
37.	90°	−5°		
38.	101°	+30′		
39.	37°	±3° 35′		
40.	83° 30′	±50′		
41.	30°	±30″		
42.	45°	+1° 15′		
43.	212° 47′	−30′		
44.	66°	±20′		

Using a protractor, draw angles of the following sizes:

45. 90°

46. 28°

47. 261°

48. 180°

49. 125°

50. 350°

Copyright by The Goodheart-Willcox Co., Inc.

154

Section 4   Measurement

51. 15°   52. 110°

53. Twelve studs are to be welded together end-to-end to form a circle. What is the angle between each
     stud? Sketch the design including the values of each angle.

54. Sixteen holes are to be drilled in a round flange. What is the angle between each hole? Sketch your
     answer including the values of each angle.

55. What size is each angle when 180° is divided into 9 equal parts?

56. A circular plate is to be divided into 15 equal parts. What is the angle of each part?

Copyright by The Goodheart-Willcox Co., Inc.

Unit 17
Four-Sided Measure

Key Terms

area	perimeter	square	trapezoid
parallelogram	rectangle	square units	

Introduction

The products that you will work on in the shop or at the job site are made up of regular geometric shapes that can be measured. Products, such as frames, braces, drums, bins, hoppers, chutes, and tanks, are examples of such items. In this unit, you will learn to calculate the perimeter and area of the most common four-sided shapes found in industry. The **perimeter** is the distance around a shape. The **area** is the surface measure of a shape.

Square

A **square** is a shape having four sides, with its opposite sides being parallel, all sides being the same length, and all its angles being 90°.

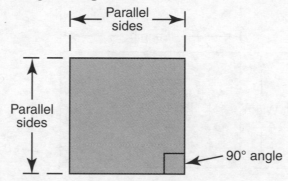

Perimeter of a Square

To calculate the perimeter of a square, add the four sides.

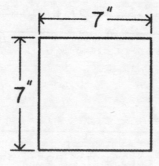

$$P = 7'' + 7'' + 7'' + 7''$$
$$P = 28''$$

Copyright by The Goodheart-Willcox Co., Inc.

Area of a Square

To calculate the area of a square, multiply the two sides.

$$A = 7'' \times 7''$$
$$A = 49 \text{ in}^2$$

Area is always expressed in **square units**, such as square feet, square millimeters, square miles, etc. Remember the rules for calculating denominate numbers. When two denominate numbers are multiplied together, their units combine. When those two units are the same, they become a squared unit, as seen in the example above. Review the principles of denominate numbers in Units 1–5. Common notations for area include the following:

49 square feet
49 sq/ft
49 ft^2
49 sq. ft.
750 mm^2

Rectangle

A **rectangle** is a shape having four sides, with its opposite sides being parallel and equal in length, and all its angles being 90°. It is similar to a square except one side is longer (length) than the adjacent side (width).

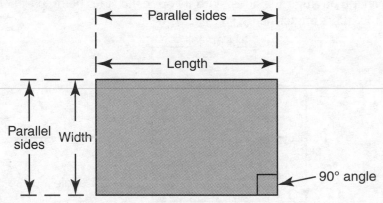

Perimeter of a Rectangle

To calculate the perimeter of a rectangle, add the four sides.

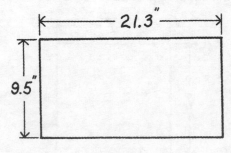

$$P = 21.3'' + 21.3'' + 9.5'' + 9.5''$$
$$P = 61.6''$$

Copyright by The Goodheart-Willcox Co., Inc.

Area of a Rectangle

To calculate the area of a rectangle, multiply length by width.

$$A = 21.3'' \times 9.5''$$

$$A = 202.35 \text{ in}^2$$

Parallelogram

A **parallelogram** is a shape having four sides, its opposite sides being the same length and are parallel. The angles opposite each other are equal. It looks like a rectangle that has been tilted.

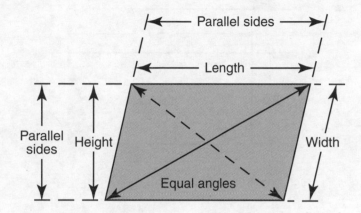

Perimeter of a Parallelogram

To calculate the perimeter of a parallelogram, add the four sides.

$$P = 193 \text{ mm} + 193 \text{ mm} + 64 \text{ mm} + 64 \text{ mm}$$

$$P = 514 \text{ mm}$$

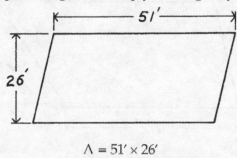

Area of a Parallelogram

To calculate the area of a parallelogram, multiply the length by the height.

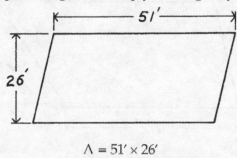

$$A = 51' \times 26'$$

$$A = 1,326 \text{ ft}^2$$

Copyright by The Goodheart-Willcox Co., Inc.

Notice it is <u>not</u> the length and the width that are being multiplied, but the length of one side and the height. You may find it interesting to note that geometric shapes are often closely related. If you were to cut the parallelogram below along the dotted line and then reassemble it as shown, you would have created a rectangle. This is why the area of a parallelogram is calculated by multiplying the length and the height.

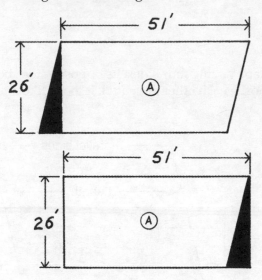

Trapezoid

The **trapezoid**, like all the shapes studied so far, has four sides. Two of the sides are parallel, but the other two sides are not parallel to each other. For this reason, a trapezoid is not as symmetric or pleasing to look at as a parallelogram. Also, it does not appear in fabrication and construction as often as the parallelogram.

Perimeter of a Trapezoid

To calculate the perimeter of a trapezoid, add the four sides.

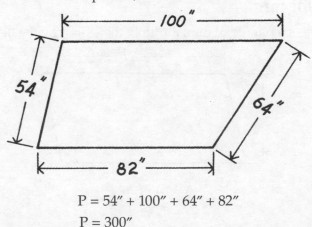

$$P = 54'' + 100'' + 64'' + 82''$$
$$P = 300''$$

Copyright by The Goodheart-Willcox Co., Inc.

Area of a Trapezoid

Use the following steps to calculate the area of a trapezoid.

1. Add the two parallel lengths.

$$331 + 262 = 593$$

2. Multiply by the height.

$$593 \times 204 = 120972$$

3. Divide by 2.

$$120972 \div 2 = 60486 \text{ mm}$$

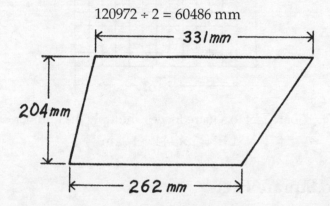

As stated previously, geometric shapes are often related. If you make two trapezoids of the same size, rotate one of them 180°, and then butt them end to end, you will have formed a parallelogram that is twice the size of the original trapezoid.

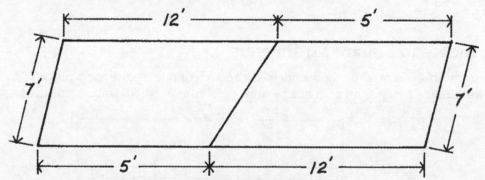

Converting Square Units of Measure

In the welding trade, the most common units of square measure are square feet, square inches, and square millimeters. Each unit of measure can be converted to the other units.

Copyright by The Goodheart-Willcox Co., Inc.

Square Feet to Square Inches

A square with sides measuring 1′ has an area of one square foot. Since 1′ equals 12″, the square also has an area of 12″ × 12″ or 144 in^2.

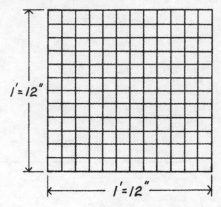

Therefore, to change square feet to square inches, multiply the number of square feet by 144.

$$1 \text{ ft}^2 = 1 \times 144 = 144 \text{ in}^2$$

Square Inches to Square Feet

To convert square inches to square feet, divide the number of square inches by 144.

1. 144 in^2 = 144 ÷ 144 = 1 ft^2
2. 1,000 in^2 = 1,000 ÷ 144 = 6.9 ft^2 (rounded)
3. 3,600 in^2 = 3,600 ÷ 144 = 25 ft^2

Square Inches to Square Millimeters

A square with sides measuring one inch has an area of one square inch. Since 1″ equals 25.4 mm, the square also has an area of 25.4 mm × 25.4 mm or 645.16 square millimeters (mm^2).

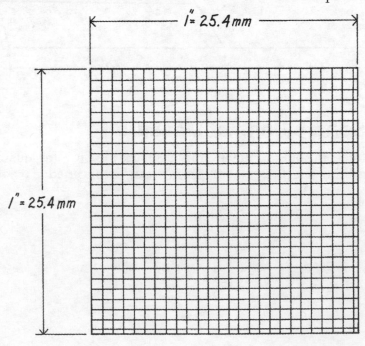

Copyright by The Goodheart-Willcox Co., Inc.

Therefore, to change square inches to square millimeters, multiply the number of square inches by 645.16.

1. 1 in^2 = 1 × 645.16 = 645.16 mm^2
2. 1,000 in^2 = 1,000 × 645.16 = 645160 mm^2
3. 2,304 in^2 = 2,304 × 645.16 = 1486449 mm^2 (rounded)

Square Millimeters to Square Inches

To convert square millimeters to square inches, divide the number of square millimeters by 645.16.

1. 645.16 mm^2 = 645.16 ÷ 645.16 = 1 in^2
2. 6000 mm^2 = 6000 ÷ 645.16 = 9.3 in^2 (rounded)
3. 500000 mm^2 = 500000 ÷ 645.16 = 775 in^2 (rounded)

From	To	Do
Square feet	Square inches	× 144
Square inches	Square feet	÷ 144
Square inches	Square millimeters	× 645.16
Square millimeters	Square inches	÷ 645.16

Copyright by The Goodheart-Willcox Co., Inc.

Work Space

Copyright by The Goodheart-Willcox Co., Inc.

Name _____ Date _____ Class _____

Unit 17 Practice

Convert the following square feet to square inches. Round to the nearest square inch. Show all your work. Be certain the columns line up. Box your answers.

1. 35 ft²

2. 115½ ft²

3. 9.72 ft²

4. 3,964 ft²

5. 0.45 ft²

6. 1,000 ft²

Convert the following square inches to square feet (to the nearest tenth). Show all your work. Be certain the columns line up. Box your answers.

7. 144 in²

8. 256¾ in²

9. 15,984 in²

10. 117.36 in²

11. 100,000 in²

12. 72 in²

Convert the following square inches to square millimeters. Round to the nearest millimeter. Show all your work. Be certain the columns line up. Box your answers.

13. 90 in²

14. 654.8 in²

15. 137⅛ in²

16. 1,008 in²

17. 1.0 in²

18. 15.5 in²

Copyright by The Goodheart-Willcox Co., Inc.

Convert the following square millimeters to square inches. Round to the nearest tenth. Show all your work. Be certain the columns line up. Box your answers.

19. 1550 mm²   20. 5161.28 mm²

21. 100 mm²   22. 9675.5 mm²

23. 645.16 mm²   24. 500000 mm²

25. A customer orders 32 pieces of the plate in the diagram below. What is the total area of the plates?

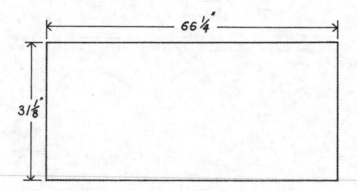

26. Calculate the perimeter of the following L-shaped plate.

27. Calculate the area of the following L-shaped plate.

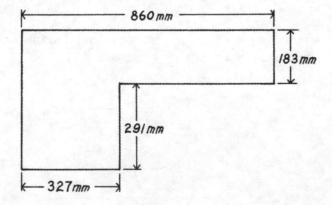

Copyright by The Goodheart-Willcox Co., Inc.

Name _____ Date _____ Class _____

28. Calculate the area of the shaded part of the figure below in square feet.

29. Convert the area of the shaded part of the figure below from square feet to square inches.

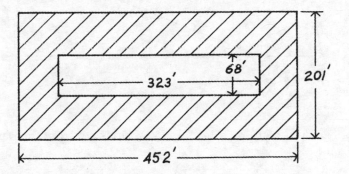

30. Calculate the perimeter of the following parallelogram

31. Calculate the area of the following parallelogram.

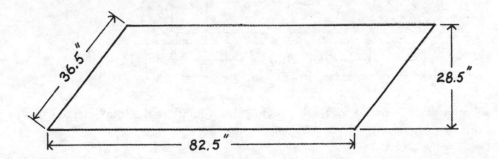

32. Calculate the perimeter of the following trapezoid.

33. Calculate the area of the following trapezoid.

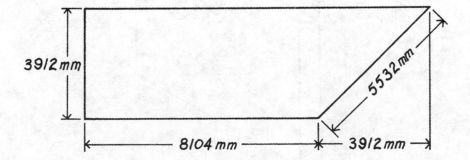

Copyright by The Goodheart-Willcox Co., Inc.

34. Calculate the area of the following trapezoid.

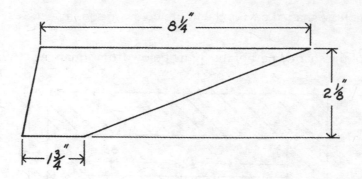

35. Calculate the area of scrap remaining after the trapezoid is cut from the parallelogram.

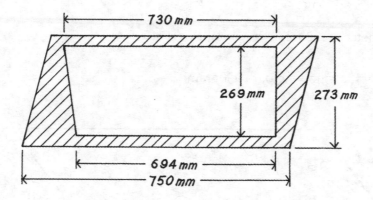

36. A heat exchanger has 16 panels of the following size. Calculate the total surface area of the panels. (Think about this problem.)

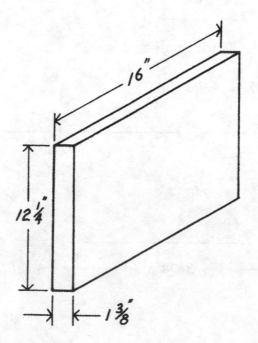

Copyright by The Goodheart-Willcox Co., Inc.

Name _____ Date _____ Class _____

37. The following plate consists of a trapezoid and a parallelogram. Calculate the perimeter of the plate.

38. Calculate the area of the plate.

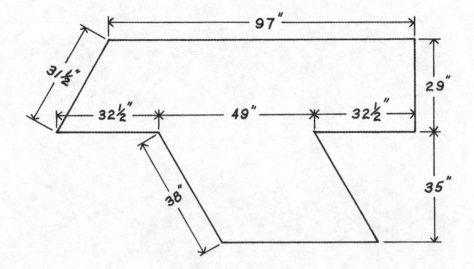

39. All surfaces of this sheet metal trough are to be painted with a rust-inhibiting paint. What is the total surface area, including the bottom, to be painted?

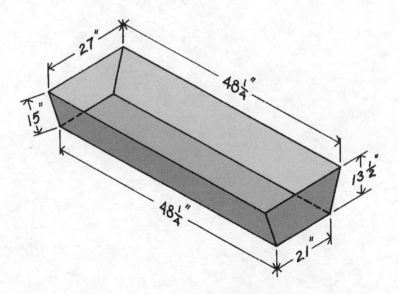

Copyright by The Goodheart-Willcox Co., Inc.

Work Space

Copyright by The Goodheart-Willcox Co., Inc.

Unit 18
Triangular Measure

Key Terms

3-4-5 triangle   hypotenuse   right triangle

equilateral triangle   isosceles triangle   triangle

Introduction

The shape explained in this unit is the three-sided figure called the **triangle**. Triangles can be divided into a number of types, but this study will be limited to the three types of triangles that will appear most frequently in your welding work: right, equilateral, and isosceles.

Right Triangle

A **right triangle** is a triangle in which one angle is 90°.

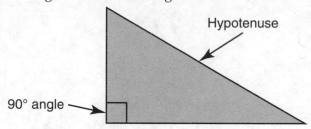

The sloping line is called the **hypotenuse**, and the other two sides are referred to as sides. The hypotenuse is always facing the 90° angle. It is never adjacent to the 90° angle.

Laying out a right angle is a job you may be called upon to do in the shop or on the job site. One method is to form a **3-4-5 triangle**. A triangle with hypotenuse and sides of these lengths will produce a right angle triangle. Any multiples of these numbers will work, such as 6-8-10, or 9-12-15, or 30-40-50, etc. The longest length will, of course, be the hypotenuse.

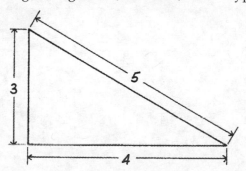

Another method of laying out a right angle is to select two pieces of the same length for the sides, say 10'. To calculate the hypotenuse, multiply the length of one side by 1.414. In this example, the hypotenuse would be 14.14'. This hypotenuse, when assembled with the two sides, will produce a right triangle.

Copyright by The Goodheart-Willcox Co., Inc.

Equilateral Triangle

An **equilateral triangle** is a triangle in which all three sides are the same length and all three angles are equal to 60°. An equilateral triangle is one of the strongest shapes in nature and is often used in fabricating.

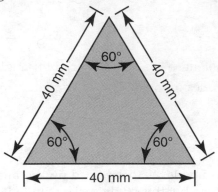

Isosceles Triangle

An **isosceles triangle** is a triangle in which two of the three sides are of equal length and two of the three angles are equal.

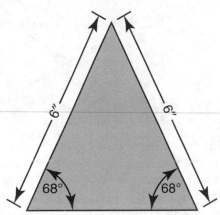

Copyright by The Goodheart-Willcox Co., Inc.

Perimeter of a Triangle

To calculate the perimeter of each of these three types of triangles, simply add the three lengths or sides.

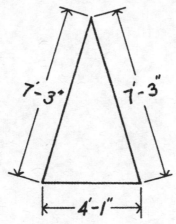

$$P = 7\text{'-}3'' + 7\text{'-}3'' + 4\text{'-}1''$$
$$P = 18\text{'-}7''$$

Area of a Triangle

To calculate the area of each of the three types of triangles, multiply ½ times the base line times the height of the triangle.

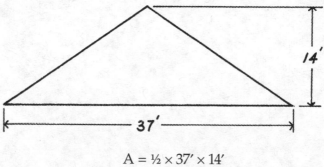

$$A = \tfrac{1}{2} \times 37' \times 14'$$
$$A = 259 \text{ ft}^2$$

It is interesting to note that the total of the three angles within any triangle is equal to 180°. So, if you know two of the angles, you can easily calculate the third.

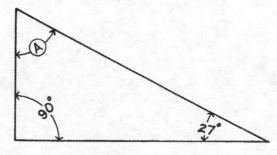

$$90° + 27° = 117°$$
$$\text{Angle A} = 180° - 117°$$
$$A = 63°$$

Copyright by The Goodheart-Willcox Co., Inc.

Work Space

Copyright by The Goodheart-Willcox Co., Inc.

Name _____ Date _____ Class _____

Unit 18 Practice

Perform the following equations as directed. Show all your work. Box your answers.

1. Calculate the perimeter of this right triangle.

2. Calculate the area of this right triangle.

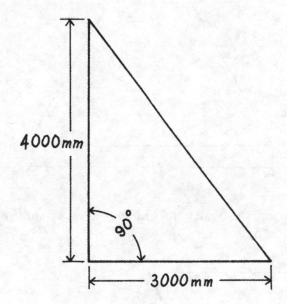

3. Calculate the perimeter of this isosceles triangle.

4. Calculate the area of this isosceles triangle.

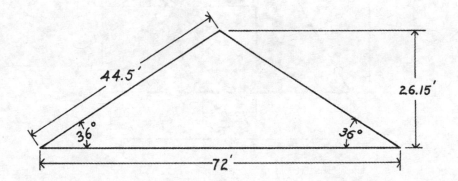

Copyright by The Goodheart-Willcox Co., Inc.

5. Calculate the perimeter of this equilateral triangle.

6. Calculate the area of this equilateral triangle.

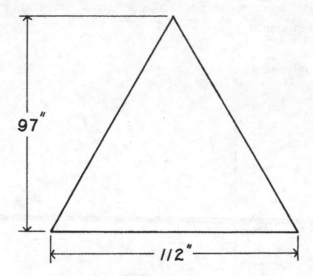

7. A rectangular hole is cut from the equilateral triangle below. What is the remaining area of the triangle?

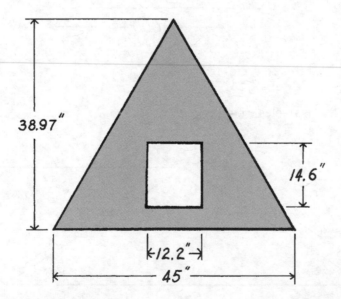

Copyright by The Goodheart-Willcox Co., Inc.

Name _____   Date _____ Class _____

8. What is distance Ⓐ in the triangle below?

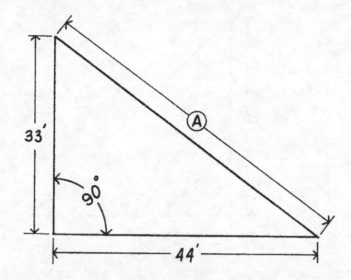

9. Calculate the total area of these nine equilateral triangles. Round to the tenth place.

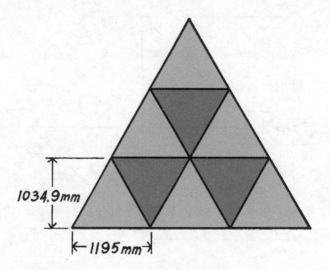

Copyright by The Goodheart-Willcox Co., Inc.

10. Calculate the perimeter of the following triangle.

11. Calculate the area of the following triangle.

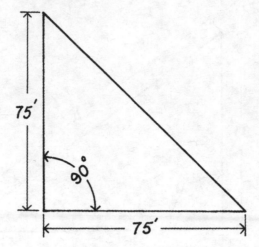

12. Review the diagram below. The centers of two bars of equal length cross at 90°. Four additional bars frame the two cross pieces. What is the total length of bar used in this piece? Express your final answer in feet and inches and round your answer to the nearest inch.

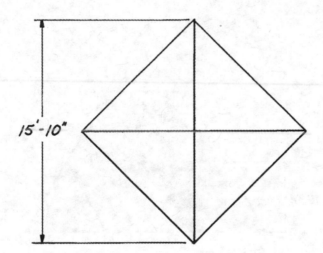

Copyright by The Goodheart-Willcox Co., Inc.

Name _____ Date _____ Class _____

13. Review the diagram below. What is the maximum length of tubing required for this bracket? Round your answer to the nearest 0.25″.

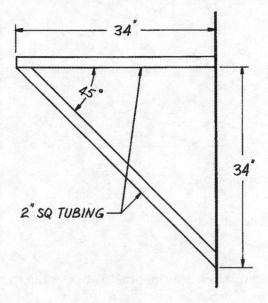

14. Calculate the combined area of the triangles (A) and (B) in the diagram below.

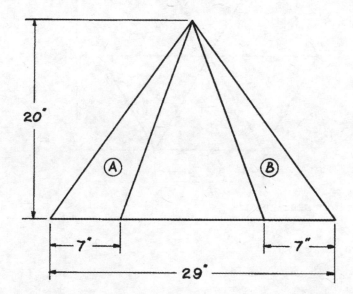

Review the diagram below and use this information to answer the two following questions. A plate measuring 48" × 102" was cut on an angle four times, resulting in the shape shown below.

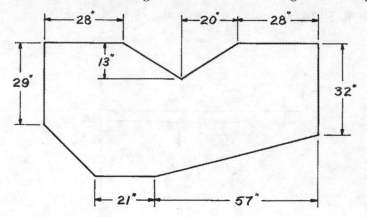

15. Calculate the original area of the plate.

16. Calculate the area of the plate after cutting.

Review the diagram below and use this information to answer the two following questions. The frame for the roof shown below is made by welding 3" S-beam.

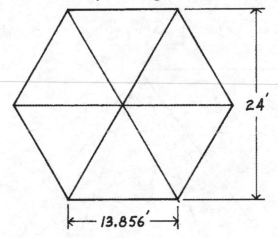

17. What is the total length of beam used?

18. What is the total area of the roof?

Copyright by The Goodheart-Willcox Co., Inc.

Name _____   Date _____   Class _____

Review the diagram of this structural frame below. Answer the two following questions.

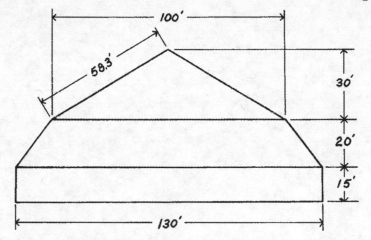

19. Calculate the total length of framing required.

20. Calculate the total area of the shape.

21. Review the diagram below. What is the total length of angle iron used in the following roof truss?

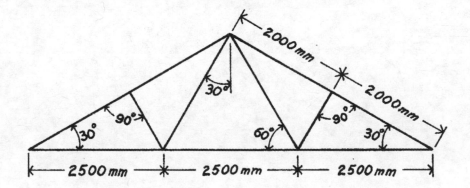

Copyright by The Goodheart-Willcox Co., Inc.

Work Space

Copyright by The Goodheart-Willcox Co., Inc.

Unit 19
Circular Measure

Key Terms

circumference   radius

diameter

Introduction

Welding operations deal with objects in a variety of shapes. Working with circles and circular figures will present no problem to the welder familiar with the basic mathematical characteristics of the circle.

Circle Terms and Principles

To begin, here is the terminology you will need to know:

- The **circumference** is the distance around a circle.
- The **diameter** is the straight line distance across a circle and passing through the center.
- The **radius** is the straight line distance from the center to the edge of a circle.

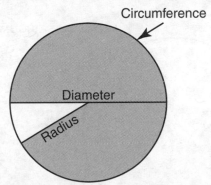

If you know certain characteristics of a circle, you can calculate other characteristics.

Circumference of a Circle

Circumference can be thought of as the perimeter of a circle. If you know the diameter of a circle, the circumference can be calculated from the following formula:

$$\text{Circumference} = \pi \times \text{Diameter}$$

$$C = \pi \times D$$

This formula requires some explanation. The symbol π is actually a letter from the Greek language, and it represents the number 3.14. In English, it is spelled "pi," but it is pronounced "pie." By multiplying the diameter of a circle by this number, you will arrive at the circumference.

So, the diameter and circumference are related to each other through the number π.

Copyright by The Goodheart-Willcox Co., Inc.

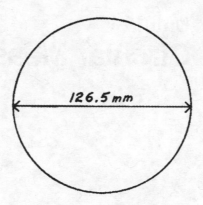

$$\text{Circumference} = \pi \times D$$
$$C = 3.14 \times 126.5 \text{ mm}$$
$$C = 397.21 \text{ mm}$$

Diameter of a Circle

If the circumference is known, the diameter of a circle can be calculated from the following formula:

$$D = \frac{C}{\pi}$$

$$\text{Diameter} = \frac{\text{Circumference}}{\pi}$$

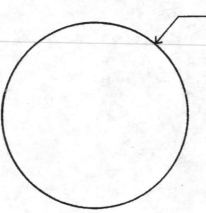

CIRCUMFERENCE = 37'

$$\text{Diameter} = \frac{C}{\pi}$$

$$D = \frac{37'}{3.14}$$

$$D = 11.8' \text{ (rounded)}$$

Area of a Circle

The area of a circle can be calculated if the radius is known.

$$\text{Area} = \pi \times \text{Radius} \times \text{Radius}$$
$$A = \pi \times R \times R$$

Copyright by The Goodheart-Willcox Co., Inc.

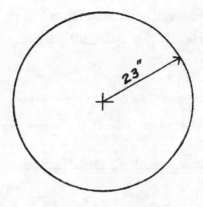

$$\text{Area} = \pi \times R \times R$$

$$A = 3.14 \times 23 \times 23$$

$$A = 1{,}661 \text{ in}^2 \text{ (rounded)}$$

The formulas for circumference, diameter, and area are extremely important, and you should memorize them.

Circular Shapes

Many fabricated objects consist of partial circles and straight lines. By recognizing basic shapes within these objects, some important characteristics can be calculated. These three examples of a half circle, a cylinder, and a semicircular sided shape shown here will give you some guidance.

Half Circle

Given the information in the diagram below, the perimeter and area can be calculated.

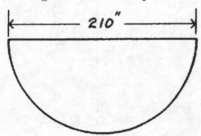

Perimeter of a Half Circle

Examining a half circle reveals that its perimeter consists of a circle's diameter and half of a circle's circumference. Knowing only diameter, its perimeter can be calculated. This is done by adding diameter and half of a circle's circumference.

$$\text{Perimeter} = D + \frac{C}{2}$$

In the example above, we first need to calculate the circumference of the circle. Then, divide that value in half to get the circumference of a half circle.

$$\text{Circumference (half circle)} = \pi \times \frac{D}{2}$$

$$C = \frac{3.14 \times 210''}{2} = 329.7''$$

Copyright by The Goodheart-Willcox Co., Inc.

Now we can use the circumference value in the perimeter formula.

Perimeter (half circle) = 329.7″ + 210″ = 539.7″

Area of a Half Circle

The area of a half circle is calculated by dividing in half the area of a whole circle.

$$\text{Area (half circle)} = \frac{\pi \times R \times R}{2}$$

Again, we will use the example above with the 210″ diameter. Since a radius is simply half a circle's diameter, this radius is 105″.

$$A = \frac{3.14 \times 105″ \times 105″}{2}$$

$$A = \frac{34{,}618.5 \text{ in}^2}{2}$$

$$\text{Area (half circle)} = 17{,}309.25 \text{ in}^2$$

Cylinder

Given the information in the diagram below, the entire area of the surface of a cylinder can be calculated. This process is broken into parts. First, calculate the area of the curved surface. Next, calculate the area of the top and bottom of the cylinder. Then, add those values to the area of the curved surface of the cylinder.

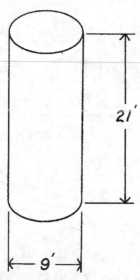

Area of a Curved Surface

The curved surface of a cylinder is deceptively simple. In reality, it is merely a rectangle wrapped as a circle. The width of the rectangle is the height of the cylinder. The length of the rectangle is the circumference of the circular end of the cylinder. Therefore, we calculate the circumference, redraw the curved surface as a flat rectangle, and use the area of a rectangle formula: Length × Width.

$$\text{Circumference} = \pi \times \text{Diameter}$$

$$C = 3.14 \times 9' = 28.26'$$

Copyright by The Goodheart-Willcox Co., Inc.

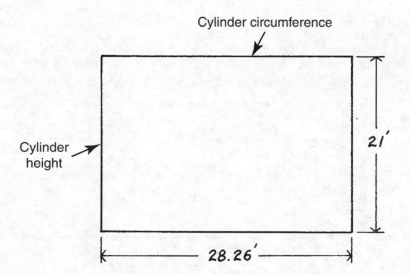

Now we have all the numbers we need to calculate the area of the curved surface of the cylinder. Plug these into the area formula.

$$\text{Area (rectangle)} = \text{Length} \times \text{Width}$$
$$\text{Area} = 21' \times 28.26' = 593.46 \text{ ft}^2$$

Area of a Cylinder

To calculate the area of the entire surface of a cylinder, add the area of the two circular ends to the area of the curved surface. Begin by calculating the area of one of the circular ends. Use the area of a circle formula.

$$\text{Area} = \pi \times \text{Radius} \times \text{Radius}$$

Since the radius is half the diameter, divide the diameter of 9' in half to get 4.5'.

$$A = 3.14 \times 4.5' \times 4.5'$$
$$A = 63.585 \text{ ft}^2$$

Since a cylinder has two circular ends, double the area value calculated above.

$$\text{Area of cylinder ends} = 63.585 \text{ ft}^2 \times 2 = 127.17 \text{ ft}^2$$

Add the area of the curved surface of the cylinder to the area of the cylinder ends to get the area of the entire surface of a cylinder.

$$\text{Area of the entire surface of a cylinder} = \text{Area of the curved surface} + \text{Area of the cylinder ends}$$
$$\text{Area} = 593.46 \text{ ft}^2 + 127.17 \text{ ft}^2 = 720.63 \text{ ft}^2$$

Copyright by The Goodheart-Willcox Co., Inc.

Semicircular Sided Shape

Semicircular sided shapes should be examined to see of which basic shapes they are composed. This information will provide clues about what formulas they require. In the diagram below, enough information is present for perimeter and area to be calculated.

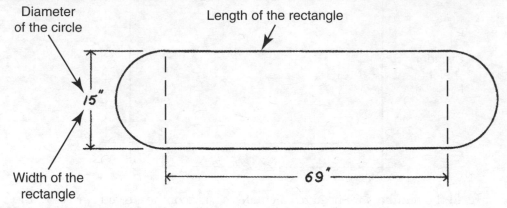

Perimeter of a Semicircular Sided Shape

Calculate the perimeter by adding the two straight sides and the semicircular ends. Since the two ends together equal one circle, use the equation for a circle's circumference. Note that the height of the shape is the same distance as the circle's diameter. These calculations are as follows:

$$\text{Circumference} = \pi \times \text{Diameter}$$
$$C = 3.14 \times 15'' = 47.1''$$

Having calculated the circumference, plug in that number to the perimeter equation.

$$\text{Perimeter} = \text{Length} + \text{Length} + \text{Circumference}$$
$$\text{Perimeter} = 69'' + 69'' + 47.1'' = 185.1''$$

Area of a Semicircular Sided Shape

The area consists of two semicircles (one complete circle) plus a rectangle in the center. Begin with the area of the circle. Since this formula requires the value of the radius, calculate that number by dividing the diameter in half.

$$\text{Radius} = \frac{\text{Diameter}}{2}$$

$$\text{Radius} = \frac{15''}{2} = 7.5''$$

Use this value of the radius in the area of a circle formula.

$$\text{Area of a circle} = \pi \times R \times R$$
$$\text{Area of a circle} = 3.14 \times 7.5'' \times 7.5''$$
$$\text{Area of a circle} = 176.625 \text{ in}^2$$

Next, calculate the area of the rectangle.

$$\text{Area of rectangle} = \text{Length} \times \text{Width}$$
$$\text{Area of rectangle} = 69'' \times 15'' = 1{,}035 \text{ in}^2$$
$$\text{Area of semicircular sided shape} = 176.625 \text{ in}^2 + 1{,}035 \text{ in}^2 = 1{,}211.625 \text{ in}^2$$

Copyright by The Goodheart-Willcox Co., Inc.

Name _____ Date _____ Class _____

Unit 19 Practice

Show all your work. Box your answers.

1. Review the diagram below. Calculate the area of scrap metal remaining after the two circle parts are cut from this piece of sheet metal. Answers should be rounded to the nearest inch.

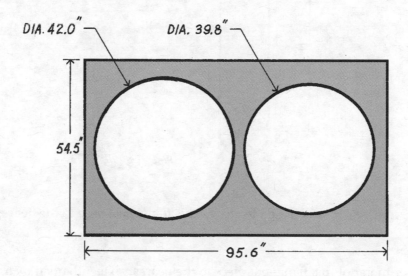

2. Review the diagram below. Calculate the area to the nearest square foot.

3. Calculate the perimeter to the nearest inch.

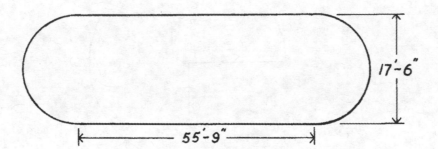

Copyright by The Goodheart-Willcox Co., Inc.

4. Review the diagram below. Calculate the length of strapping needed for this pipe hanger to the nearest sixteenth of an inch.

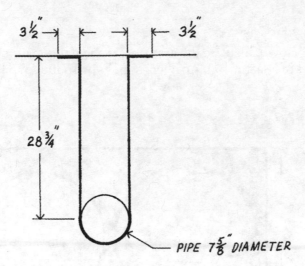

5. Review the diagram below. Calculate the area of this flame-cut piece to the nearest sixteenth of an inch.

6. Calculate the perimeter of this flame-cut piece to the nearest sixteenth of an inch.

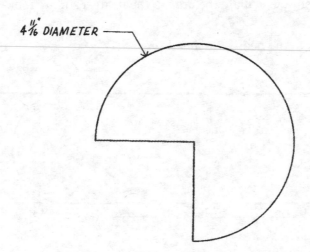

Copyright by The Goodheart-Willcox Co., Inc.

Name _____ Date _____ Class _____

7. Review the diagram below. Calculate the area of the plate before the holes have been cut. Calculate to the nearest inch.

8. Calculate the area of the plate after the holes have been cut. Calculate to the nearest inch.

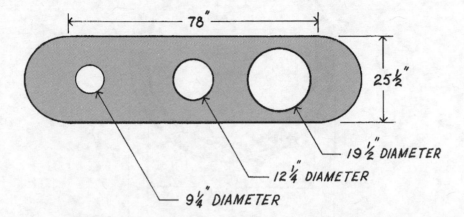

9. Review the diagram below. Calculate the total area of this cylinder to the nearest inch.

10. Convert your answer from the previous question to the nearest millimeter value.

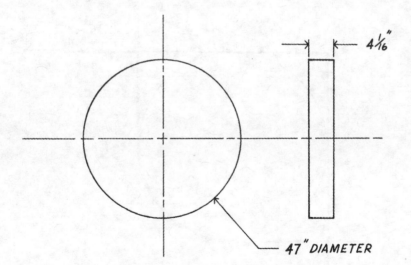

Copyright by The Goodheart-Willcox Co., Inc.

11. Calculate the area of this steel ring to the nearest mm.

12. Convert your answer from the previous question to the nearest tenth of a square inch.

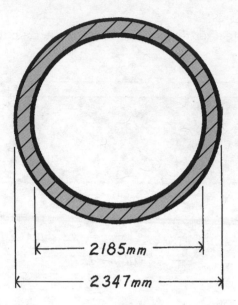

13. Review the diagram below. Calculate the total length to the nearest tenth of an inch of strapping needed for 15 pipe corner braces.

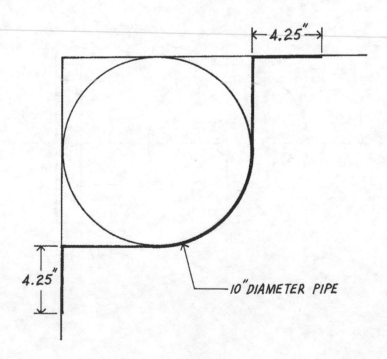

Copyright by The Goodheart-Willcox Co., Inc.

Name _____ Date _____ Class _____

14. Review the diagram below. Calculate the total length of curved pieces needed for 39 of these wrought iron sections.

15. Calculate the total length of straight pieces needed for 39 of these wrought iron sections.

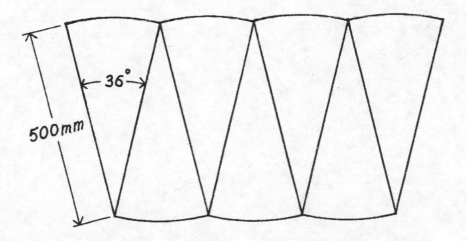

16. Review the diagram below. Find the area of this shape to the nearest inch.

17. Find the perimeter of this shape to the nearest inch.

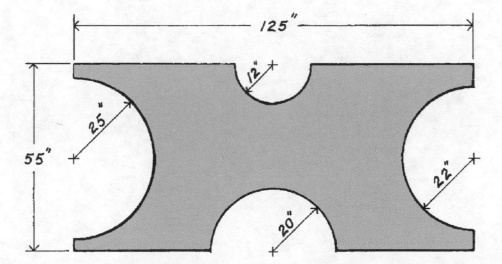

Copyright by The Goodheart-Willcox Co., Inc.

Work Space

Copyright by The Goodheart-Willcox Co., Inc.

Section 5

Volume, Weight, and Bending Metal

Section Objectives

After studying this section, you will be able to:

- Define regular shaped objects
- Calculate the volume of regular shaped objects
- Convert volume measurements between units
- Calculate the volume of irregular shaped objects
- Compute weight calculations based on volume
- Compute weight calculations based on length
- Define four types of metal bends
- Define inside and outside corner
- Compute stock length for various metal bends

Copyright by The Goodheart-Willcox Co., Inc.

Unit 20
Volume Measure

Key Terms

liter   volume

Introduction

Volume is the amount of space an object occupies. All objects exist in three dimensions: length, width, and height. The volume of an object is the product of these dimensions and is expressed in cubic units. For example:

82 cubic inches

119 cubic feet

1789 mm^3

37.5 ft^3

10 in^3

Volume of Regular Shaped Objects

The volume of an object having a uniform cross section is determined by calculating the area of the cross section and then multiplying by the length.

$$Volume = Area \times Length$$
$$V = A \times L$$

Cube

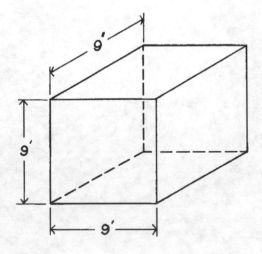

Area of cross section: 9″ × 9″ = 81 in^2

Volume: 81 in^2 × 9″ = 729 ft^3

Copyright by The Goodheart-Willcox Co., Inc.

Solid Rectangle

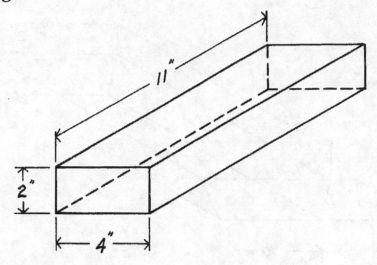

Area of cross section: $2'' \times 4'' = 8 \text{ in}^2$

Volume: $8 \text{ in}^2 \times 11'' = 88 \text{ in}^3$

Solid Parallelogram

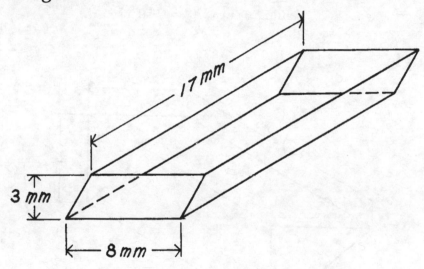

Area of cross section: $3 \text{ mm} \times 8 \text{ mm} = 24 \text{ mm}^2$

Volume: $24 \text{ mm}^2 \times 17 \text{ mm} = 408 \text{ mm}^3$

Copyright by The Goodheart-Willcox Co., Inc.

Solid Trapezoid

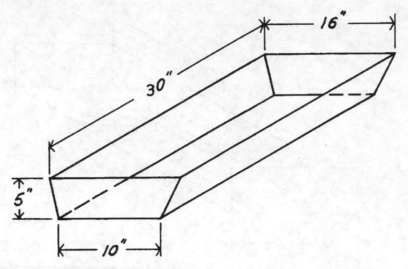

Area of cross section: $16'' + 10'' = 26''$

$26'' \times 5'' = 130 \text{ in}^2$

$130 \text{ in}^2 \div 2 = 65 \text{ in}^2$

Volume: $65 \text{ in}^2 \times 30'' = 1{,}950 \text{ in}^3$

Solid Triangle

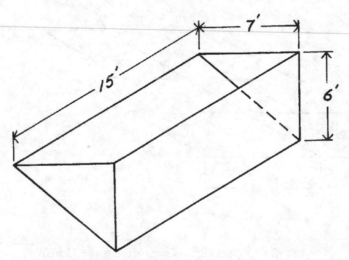

Area of cross section: $7' \times 6' \times ½ = 21 \text{ in}^2$

Volume: $21 \text{ in}^2 \times 15' = 315 \text{ ft}^3$

Copyright by The Goodheart-Willcox Co., Inc.

Cylinder

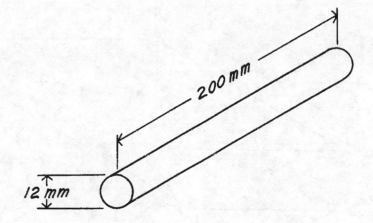

Area of cross section: 3.14 × 6 mm × 6 mm = 113.04 mm²

Volume: 113.04 mm² × 200 mm = 22608 mm³

Solid Half Circle

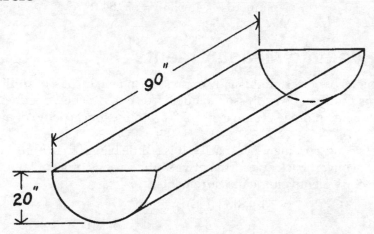

Area of cross section: $\dfrac{3.14 \times 20'' \times 20''}{2} = 628 \text{ in}^2$

Volume: 628 in² × 90″ = 56,520 in³

Copyright by The Goodheart-Willcox Co., Inc.

Solid Semicircular Sided Shape

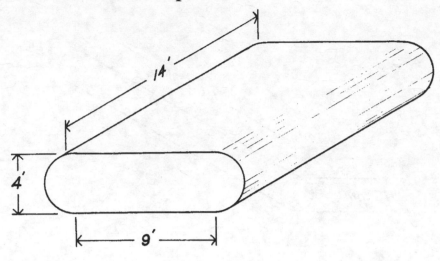

Area of cross section: 3.14 $\times$ $2' \times 2'$ = $12.56\ \text{ft}^2$

$9'$ $\times$ $4'$ = $36\ \text{ft}^2$

$12.56\ \text{ft}^2$ + $36\ \text{ft}^2$ = $48.56\ \text{ft}^2$

Volume: $14' \times 48.56\ \text{ft}^2 = 679.84\ \text{ft}^3$

Converting Volume Measurements

Linear dimensions of objects are usually expressed in feet, inches, or millimeters. Volume, as illustrated in this unit, can be expressed in cubic feet, cubic inches, or cubic millimeters. However, it is quite often more useful to express the volume (or capacity) of an object in gallons or liters.

A **liter** is a metric measurement of capacity. It is a little bigger than a quart. It is important for you to be able to convert back and forth between the various units of volume. Listed below are some common equivalent volume measurements:

$$1\ \text{gallon (US)} = 231\ \text{in}^3$$
$$1\ \text{ft}^3 \qquad = 7.48\ \text{gallons (US)}$$
$$1\ \text{gallon (US)} = 3.785\ \text{liters}$$
$$1\ \text{ft}^3 \qquad = 1{,}728\ \text{in}^3$$
$$1\ \text{in}^3 \qquad = 16387.064\ \text{mm}^3$$

The degree of accuracy usually required in a welding shop allows the use of rounded figures.

$$1\ \text{in}^3 \qquad = 16387\ \text{mm}^3$$
$$1\ \text{gallon (US)} = 3.79\ \text{liters}$$

Also, since conversions often result in length decimals, it is normal practice to round the answer.

1. **Gallons to Cubic Inches:**
   Multiply the number of gallons by 231.
   $40\ \text{gallons} = 40 \times 231 = 9{,}240\ \text{in}^3$

2. **Cubic Inches to Gallons:**
   Divide the number of cubic inches by 231.
   $2{,}425.5\ \text{in}^3 = 2{,}425.5 \div 231 = 10.5\ \text{gallons}$

Copyright by The Goodheart-Willcox Co., Inc.

3. **Gallons to Cubic Feet:**
Divide the number of gallons by 7.48.
1,000 gallons = 1,000 ÷ 7.48 = 133.6898 ft^3 = 133.69 ft^3 (rounded)

4. **Cubic Feet to Gallons:**
Multiply the number of cubic feet by 7.48.
16 ft^3 = 16 × 7.48 = 119.68 gallons

5. **Gallons to Liters:**
Multiply the number of gallons by 3.79.
5 gallons = 5 × 3.79 = 18.95 liters

6. **Liters to Gallons:**
Divide the number of liters by 3.79.
85 liters = 85 ÷ 3.79 = 22.4274 gallons = 22.43 gallons (rounded)

7. **Cubic Millimeters to Cubic Inches:**
Divide the number of cubic millimeters by 16,387.
27000000 mm^3 = 27000000 ÷ 16,387 = 1,647.65 in^3 (rounded)

8. **Cubic Inches to Cubic Millimeters:**
Multiply the number of cubic inches by 16,387.
9 in^3 = 9 × 16,387 = 147483 mm^3

9. **Cubic Feet to Cubic Inches:**
Multiply the number of cubic feet by 1,728.
27 ft^3 = 27 × 1,728 = 46,656 in^3

10. **Cubic Inches to Cubic Feet:**
Divide the number of cubic inches by 1,728.
5,000 in^3 = 5,000 ÷ 1,728 = 2.8935 ft^3 = 2.89 ft^3 (rounded)

Volume of Irregular Shaped Objects

The volume of many irregular shaped objects can be easily calculated. The way to compute this is to first divide the object into regular shaped parts. Then calculate the volume of each part. Finally, add all the calculated volumes. The example below calculates the volume to the nearest tenth of an inch. The material is 2″ thick and all holes are 2″ in diameter.

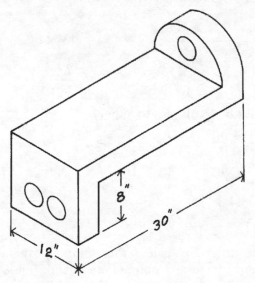

Copyright by The Goodheart-Willcox Co., Inc.

1. Step One: Divide the object into regular shaped parts.

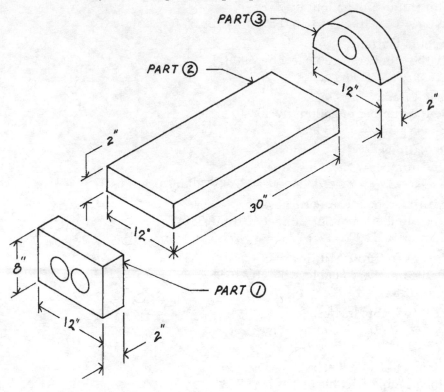

2. Step Two: Calculate the volume of each part.
    Part 1:

Gross volume = $2'' \times 8'' \times 12''$	$= \quad 192.00 \text{ in}^3$
Minus volume of holes = $\pi \times 1'' \times 1'' \times 2'' \times 2$	$= - \;\; 12.56 \text{ in}^3$
Volume	$= \quad 179.44 \text{ in}^3$

Part 2:

Volume = $2'' \times 12'' \times 30''$ =   720.00 in^3

Part 3:

$$\text{Volume of half circle} = \frac{\pi \times 6'' \times 6'' \times 2''}{2} \quad = \quad 113.04 \text{ in}^3$$

Minus volume of hole = $\pi \times 1'' \times 1'' \times 2''$	$= - \;\; 6.28 \text{ in}^3$
Volume =	106.76 in^3

3. Step Three: Add the individual volumes together.

Part 1:	179.44 in^3
Part 2:	720.00 in^3
Part 3:	+ 106.76 in^3
Total Volume:	1,006.2 in^3

A Word of Caution

One of the difficulties encountered in calculating volumes, especially of irregular shaped objects, is keeping track of your figures. As pointed out in Unit 1, it is crucial that you form the habit of being organized in your math work. You will find that math concepts do not become more difficult. However, the sheer volume of figures you will be dealing with may cause confusion. So, one of the main challenges you will continually face is to maintain order and control of your math work.

Copyright by The Goodheart-Willcox Co., Inc.

Name _____ Date _____ Class _____

Unit 20 Practice

Calculate the following equations. Show all your work. Be certain the columns line up. Box your answers.

1. How many cubic inches of space does 6¼ gallons of water occupy?

2. How many cubic feet are contained in 29000000 mm³? Calculate to the nearest tenth.

3. A swimming pool contains 1,600 ft³ of water. How many gallons does the pool contain? Calculate to the nearest gallon.

4. Convert 1.125 in³ to mm³. Calculate to the nearest mm.

5. A home heating oil storage tank holds 480 liters of oil. How many gallons does it hold? Calculate to the nearest gallon.

6. A microwave oven measures 16½″ × 11″ × 7½″. What is the volume in cubic inches? Calculate to the nearest hundredth.

7. What is the volume in cubic feet of the microwave oven from the previous question? Calculate to the nearest hundredth.

8. How many liters are there in a 25 gallon gasoline tank? Calculate to the nearest tenth.

9. How many cubic inches of space are there in a refrigerator with a capacity of 8 ft³?

10. The water tank for the town of South Windsor has a capacity of 500,000 gallons. How many cubic feet of water does it contain? Calculate to the nearest cubic foot.

11. A motorcycle engine has a size of 45 in³. What is the size of the engine in mm³?

Copyright by The Goodheart-Willcox Co., Inc.

12. Convert 9,000 in³ to cubic feet. Calculate to the nearest tenth.

13. A quenching tank measures 2′ × 3′ × 6′. How many gallons does it hold? Calculate to the nearest hundredth.

14. How many gallons are there in 1,000 in³? Calculate to the nearest thousandth.

15. A block of precision steel contains 2539985 mm³. What is the volume in cubic inches?

16. Convert 32,000 gallons to cubic feet. Calculate to the nearest tenth.

17. How many cubic millimeters are contained in the following piece of steel? Calculate to the nearest hundredth mm.

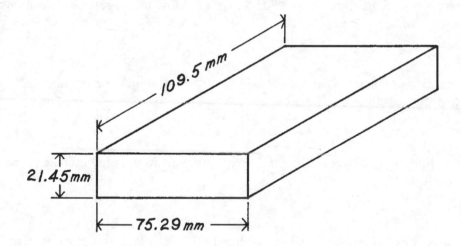

Copyright by The Goodheart-Willcox Co., Inc.

Name _____ Date _____ Class _____

18. What is the volume of steel in this triangular rod? Calculate to the nearest cubic inch.

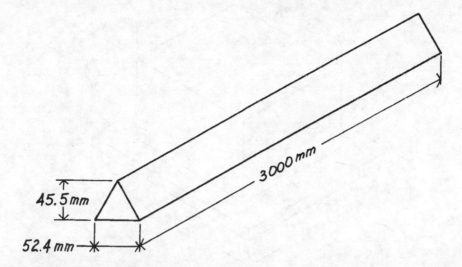

19. Review the diagram below. The concrete foundation for a newly installed press has the following dimensions. How many cubic yards of concrete were used (to the nearest one quarter of a yard)?

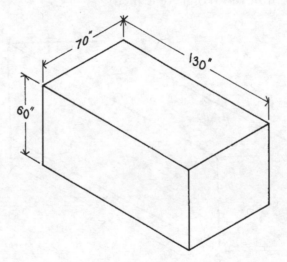

Copyright by The Goodheart-Willcox Co., Inc.

20. Find the total volume of this weldment to the nearest half cubic inch.

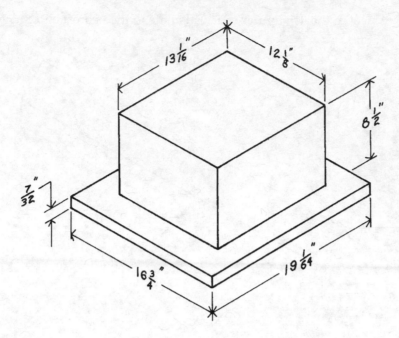

21. This block of steel is turned on a lathe to a diameter of 18″. What volume of material is removed (to the nearest hundredth of a cubic inch)?

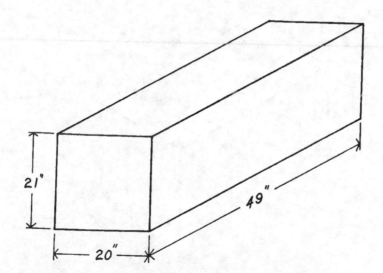

Copyright by The Goodheart-Willcox Co., Inc.

Name _____ Date _____ Class _____

22. Review the diagram below. Calculate the volume of this shape.

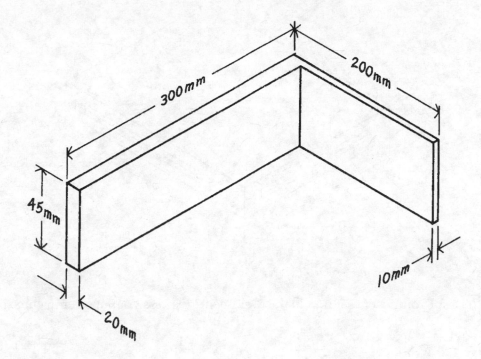

23. Fifteen joints are welded as illustrated. Fillet welds on both sides of the joint are similar. What volume of weld material is deposited?

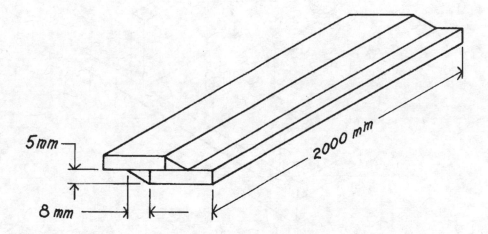

Copyright by The Goodheart-Willcox Co., Inc.

24. The ventilating system in this fabricating shop can evacuate 6,100 ft³ of air per minute. How long will it take to exchange all of the air in the shop? Calculate to the nearest minute.

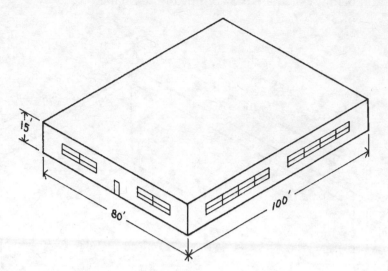

25. What is the total volume of cast iron in this weldment? Express your answer in cubic inches.

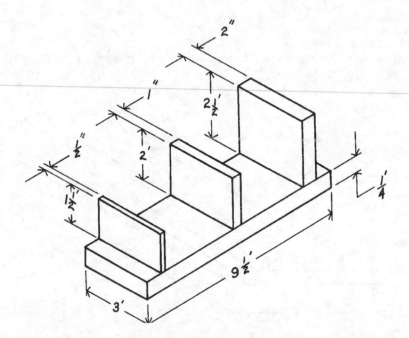

Copyright by The Goodheart-Willcox Co., Inc.

Name _____ Date _____ Class _____

26. Calculate the volume of this silo to the nearest cubic foot.

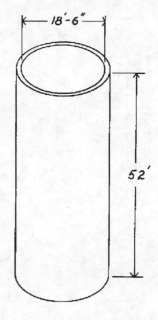

27. How many cubic feet of steel are in this 250′ roll?

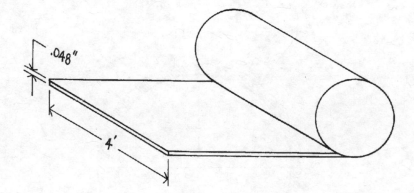

Copyright by The Goodheart-Willcox Co., Inc.

28. Nine bins are to be constructed as shown below. What is the total volume of the bins to the nearest cubic foot?

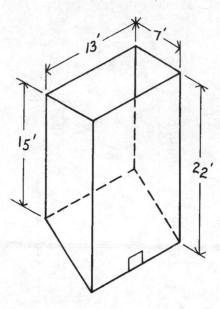

29. The worn surface of this 475 mm long shaft is built up by surfacing. Three millimeters of material are deposited all around. What total volume of weld material is deposited?

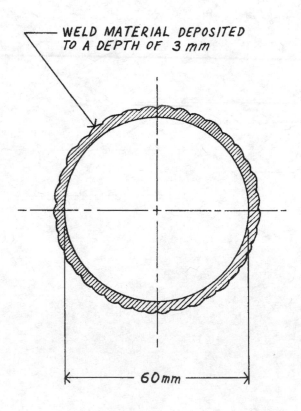

WELD MATERIAL DEPOSITED
TO A DEPTH OF 3 mm

60mm

Copyright by The Goodheart-Willcox Co., Inc.

Name _____ Date _____ Class _____

30. This 17′ wide above-ground pool is filled to within 6″ of the top. How many liters of water does it contain (to the nearest ten liters)?

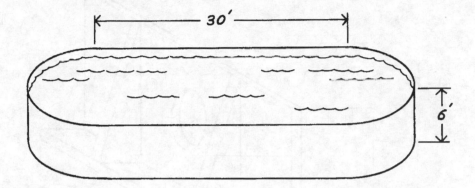

31. Review the diagram below. Which container has greater capacity: the rectangular or the half circle container?

32. How many more gallons can the container with greater capacity hold? Calculate to the nearest gallon.

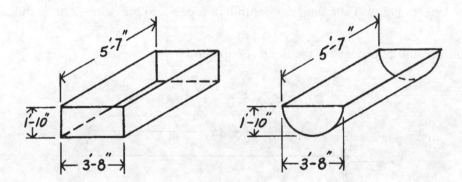

Copyright by The Goodheart-Willcox Co., Inc.

33. Review the diagram below. All material in this weldment is ¼″ thick unless otherwise shown on the drawing. The flame-cut hole is 10″ in diameter. The triangular support pieces are only on one side of the upright. Find the total volume to the nearest quarter of a cubic inch.

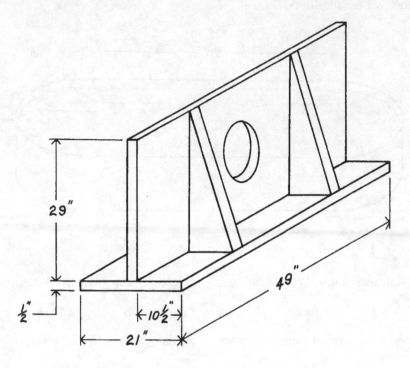

34. Review the diagram below. Calculate the volume of pipe Ⓐ to the nearest cubic inch.

35. Calculate the volume of pipe Ⓑ to the nearest cubic inch.

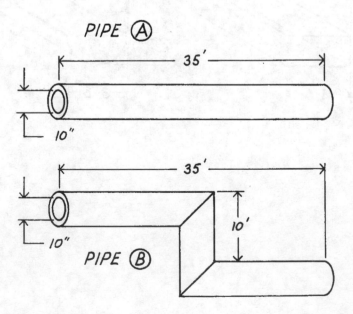

Copyright by The Goodheart-Willcox Co., Inc.

Name _____ Date _____ Class _____

36.   Review the diagram below. How many gallons will this trough hold? Calculate to the nearest
      gallon.

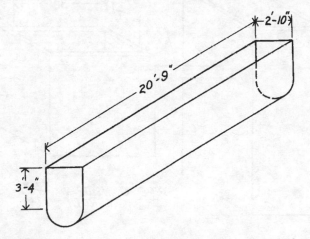

37.   Two plates, 3500 mm long, are welded together as shown. What is the total volume of weld
      deposited?

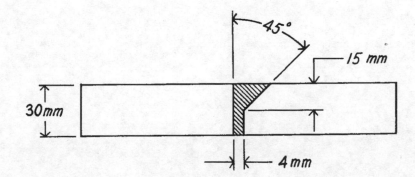

Copyright by The Goodheart-Willcox Co., Inc.

38. Calculate the total volume of the two settling tanks, including the pipes, to the nearest gallon.

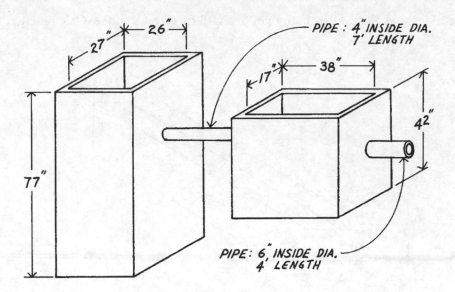

Copyright by The Goodheart-Willcox Co., Inc.

Unit 21
Weight Measure

Key Terms

C Shape	gram (g)	L Shape	S Shape
centimeter (cm)	kilogram (kg)	meter (m)	W Shape

Introduction

When calculating the weight of weldments, you will encounter one of the following conditions. The weldment may consist of randomly shaped metal plates or standard structural shapes (such as angle iron, channel, etc.) or a combination of both. The calculations for randomly shaped plates are based on volume. The calculations for standard structural shapes are based on length.

Weight Calculation Based on Volume

The weight of a block of steel is easy to determine if you know the volume of the piece and the weight of the steel per unit volume. An example will clarify this. Steel has a weight of 0.2835 lb per cubic inch; therefore, a block of steel with a volume of 250 in³ weighs 250 × 0.2835 or 70.875 lb.

Metals have different densities and, therefore, different weights, as the following list indicates. Even the weight of different classifications of steel will vary depending on the composition of the steel.

Weight of Metals					
	lb/in³	g/cm³		lb/in³	g/cm³
Magnesium	0.0628	1.738	**Steel**	0.2835	7.847
Aluminum	0.0975	2.699	**Copper**	0.3210	8.885
Zinc	0.2570	7.114	**Lead**	0.4096	11.338
Tin	0.2633	7.288	**Tungsten**	0.6900	19.099
Cast Iron	0.2665	7.377	**Gold**	0.6969	19.290

You will notice the above list introduces a new unit of measure, namely, g/cm³. This means grams per cubic centimeter. Grams (g) and centimeters (cm) are metric measurements commonly used in these calculations.

The **gram (g)** is a measure of weight and is a very small measure indeed. A paper clip, for example, weighs about one gram. Since many objects handled in everyday life are much heavier than a few grams, it is common practice to use 1000 grams as a unit of measure. One thousand grams is called a **kilogram** and is a little more than two pounds. You will see it written in two other styles, either as **kg** or as **kilo**. To convert from grams to kilograms, divide by 1000. To convert from kilograms to grams, multiply by 1000.

Copyright by The Goodheart-Willcox Co., Inc.

The **centimeter (cm)** is a measure of length and is ten millimeters long. A cube of sugar, for example, measures about 1 cm × 1 cm × 1 cm and is therefore about 1 cm³ in volume.

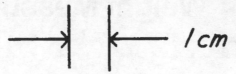

Because the millimeter is so small, many objects are measured in centimeters. To convert from millimeters to centimeters, divide by 10. To convert from centimeters to millimeters, multiply by 10.

Weight Calculation Based on Length

A full description of the standard structural shapes produced by American steel makers is provided in the *Steel Construction Manual* issued by the American Institute of Steel Construction. The Canadian equivalent is produced by the Canadian Institute of Steel Construction. Among the statistics provided for each shape is the weight per unit length. A summary of selected shapes is listed below.

You will notice that the figures may be expressed in the US Customary system as pounds per foot or in the metric system as kilograms per meter. As previously explained, 1 kilogram equals 1000 grams. Similarly, 1 **meter (m)** is equal to 1000 millimeters and is a bit longer than 1 yard. You will see it written as **m**. To convert from millimeters to meters, divide by 1000. To convert from m to mm, multiply by 1000.

The weight of a standard structural shape is easy to determine if you know the length of the piece and the weight of the shape per unit length. For example, if a 4″ channel weighs 5.4 lb per ft., the weight of a 6′ piece would be 5.4 × 6 or 32.4 lb.

Review the following standard shapes and their weights at different lengths.

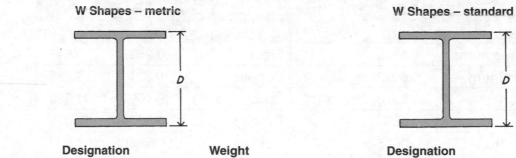

W Shapes – metric		W Shapes – standard	
Designation	**Weight**	**Designation**	**Weight**
D (mm) × weight (kg/m)	kg/m	D (inches) × weight (lb/ft)	lb/ft
W 310 × 202	202	W 16 × 57	57
W 250 × 167	167	W 16 × 26	26
W 250 × 67	67	W 12 × 35	35
W 200 × 100	100	W 8 × 28	28
W 200 × 22	22	W 6 × 25	25
W 150 × 24	24	W 6 × 9	9
W 130 × 28	28	W 5 × 19	19

Copyright by The Goodheart-Willcox Co., Inc.

S Shapes – metric

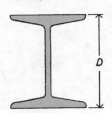

Designation D (mm) × weight (kg/m)	Weight kg/m
S 610 × 158	158
S 250 × 52	52
S 200 × 34	34
S 180 × 30	30
S 150 × 26	26
S 130 × 22	22
S 75 × 11	11

S Shapes – standard

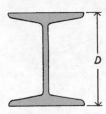

Designation D (inches) × weight (lb/ft)	Weight lb/ft
S 18 × 54.7	54.7
S 12 × 50	50.0
S 12 × 40.8	40.8
S 8 × 23	23.0
S 8 × 18.4	18.4
S 6 × 17.25	17.25
S 3 × 5.7	5.7

Channels (C Shapes) – metric

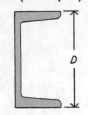

Designation D (mm) × weight (kg/m)	Weight kg/m
C 250 × 37	37
C 200 × 28	28
C 180 × 18	18
C 150 × 19	19
C 130 × 17	17
C 130 × 13	13
C 100 × 9	9

Channels (C Shapes) – standard

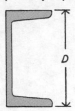

Designation D (inches) × weight (lb/ft)	Weight lb/ft
C 10 × 30	30.0
C 9 × 13.4	13.4
C 8 × 18.75	18.75
C 8 × 11.5	11.5
C 6 × 10.5	10.5
C 5 × 9	9.0
C 4 × 5.4	5.4

Angles (L Shapes) – metric

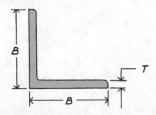

Designation B (mm) × B (mm) × T (mm)	Weight kg/m
L 200 × 200 × 16	48.216
L 150 × 100 × 13	24.257
L 125 × 90 × 13	20.685
L 100 × 100 × 16	23.685
L 90 × 75 × 10	12.173
L 55 × 55 × 8	6.399
L 45 × 55 × 6	3.244

Angles (L Shapes) – standard

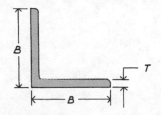

Designation B (inches) × B (inches) × T (inches)	Weight lb/ft
L 6 × 4 × ⅞	27.2
L 4 × 3½ × ½	11.9
L 3 × 3 × ⅜	7.2
L 3½ × 2½ × ⁷⁄₁₆	8.3
L 3 × 3 × ½	9.4
L 2½ × 2 × ⅜	5.3
L 2 × 2 × ⅛	1.65

Copyright by The Goodheart-Willcox Co., Inc.

Work Space

Copyright by The Goodheart-Willcox Co., Inc.

Name _____ Date _____ Class _____

Unit 21 Practice

Perform the following calculations as directed. Show all your work. Box your answers.

1. Calculate the weight of this steel bar to the nearest kilogram. (HINT: Since the table of structural steel shapes provides the weight in g/cm³, it is best to calculate the volume in cubic centimeters. First, convert the millimeter dimensions to centimeters and then proceed with finding the volume in cubic centimeters.)

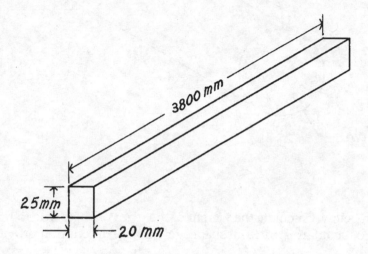

2. Review the diagram below. Calculate the weight of the copper plate to the nearest pound.

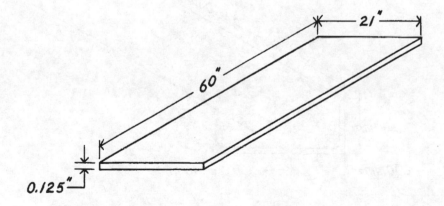

Copyright by The Goodheart-Willcox Co., Inc.

3. Review the diagram below. Calculate the weight of the cast iron bar to the nearest pound.

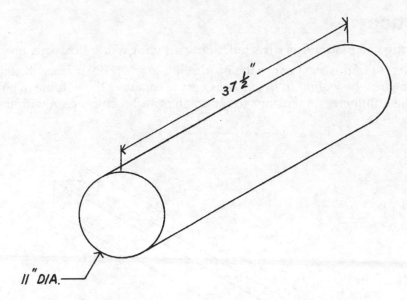

4. Review the diagram below. Calculate the weight of the piece of angle iron to the nearest tenth of a kilogram. (NOTE: For standard structural shapes refer to the description provided in the text.)

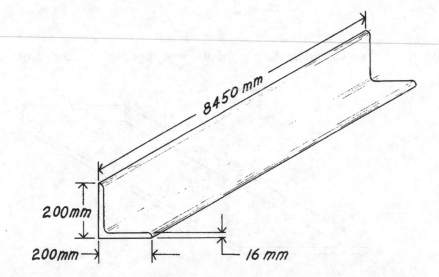

Copyright by The Goodheart-Willcox Co., Inc.

Name _____ Date _____ Class _____

5. Calculate the weight of the following S shape beam to the nearest tenth of a kilo.

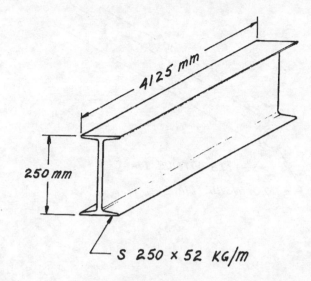

6. Calculate the weight of the following triangular piece of steel to the nearest tenth of a kilogram.

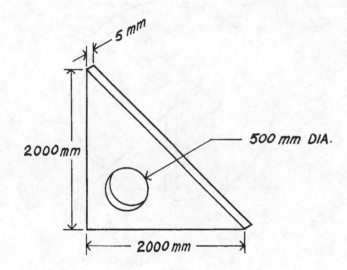

Copyright by The Goodheart-Willcox Co., Inc.

7. Calculate the weight of the following steel tube to the nearest pound.

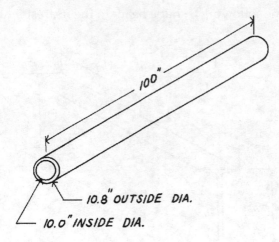

8. Calculate the weight of this piece of steel to the nearest tenth of a pound. The diameter is 40″ and the thickness is 2″.

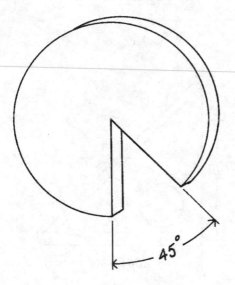

Copyright by The Goodheart-Willcox Co., Inc.

Name _____ Date _____ Class _____

9. Review the diagram of a gold bar below. Calculate the weight of this bar of gold to the nearest hundredth of a pound.

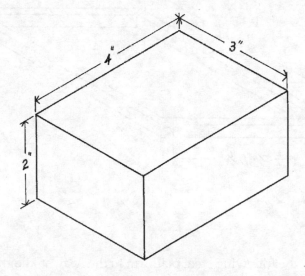

10. Calculate the weight of this steel hub and plate to the nearest tenth of a pound.

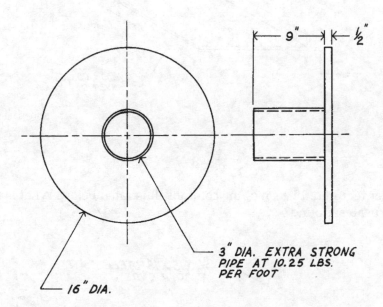

3" DIA. EXTRA STRONG
PIPE AT 10.25 LBS.
PER FOOT

16" DIA.

Copyright by The Goodheart-Willcox Co., Inc.

11. The following frame is made of square structural tubing weighing 47.9 lb per foot. What is the weight of the frame to the nearest tenth of a pound?

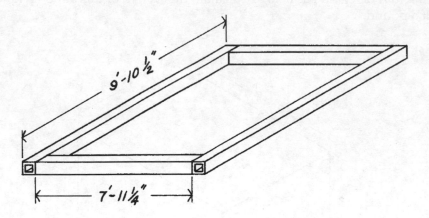

12. Calculate the weight of the following steel platform to the nearest kilogram.

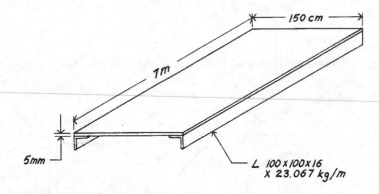

13. A customer order required thirteen of the columns illustrated below. What is the total weight of the order to the nearest pound?

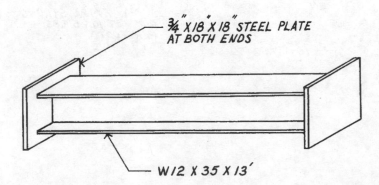

Copyright by The Goodheart-Willcox Co., Inc.

Name _____ Date _____ Class _____

14. Calculate the weight of the following weldment to the nearest kilogram.

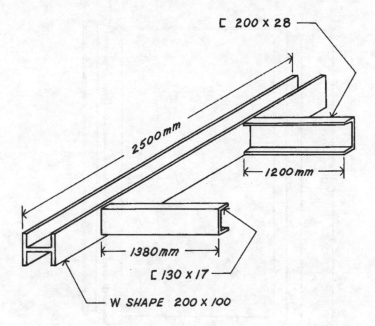

15. A machine base was fabricated from the following list of materials. What is the weight of the
    machine base to the nearest pound?

Item	Quantity	Description	Size	Weight Per Foot
A	1	Steel plate	3/8" × 25" × 40"	—
B	2	Angle iron	2" × 2" × 1/8" × 39"	1.65
C	4	Angle iron	2 1/2" × 2" × 3/8" × 13 1/2"	5.3
D	6	Steel plate	1/2" × 3" × 3"	—
E	4	Angle iron	3" × 3" × 3/8" × 17 5/8"	7.2
F	1	S shape	3" × 5.7 lb × 2'-6"	5.7
G	11	Solid round bar	1 1/8" dia. × 7"	3.38

Copyright by The Goodheart-Willcox Co., Inc.

16. Calculate the weight of the following steel pillar to the nearest pound.

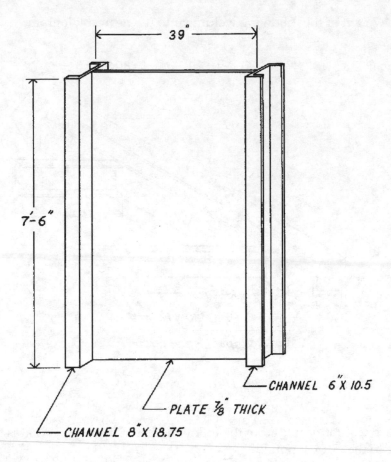

17. What is the weight of the following steel base for a swivel chair? Calculate your answer to the nearest gram.

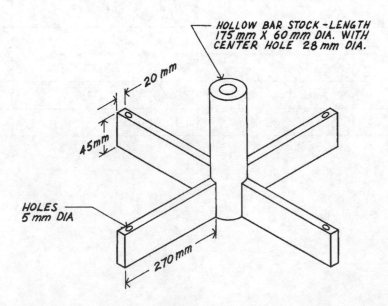

Copyright by The Goodheart-Willcox Co., Inc.

Name _____ Date _____ Class _____

18. Calculate the weight of this section of steel fence to the nearest pound.

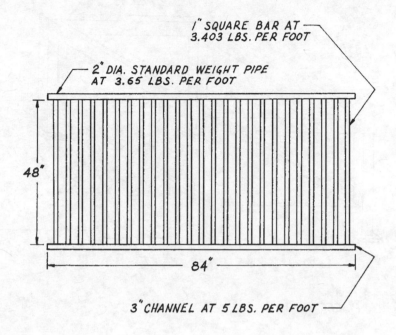

19. Calculate the weight of the following test piece to the nearest tenth of a pound. All plates are ⅜″ thick and the overall length is 38″. (HINT: A difficulty with this type of problem is keeping track of all the pieces. Try labeling the pieces as a, b, c, etc. Then, calculate the volume for each piece. Total all volumes and then calculate the total weight.)

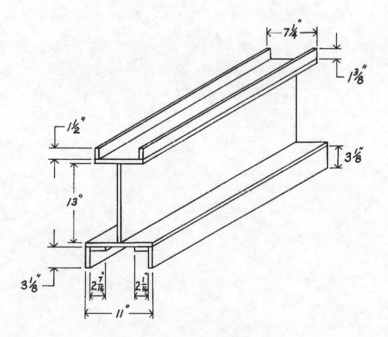

Copyright by The Goodheart-Willcox Co., Inc.

20. Calculate the weight of the following table frame to the nearest tenth of a pound.

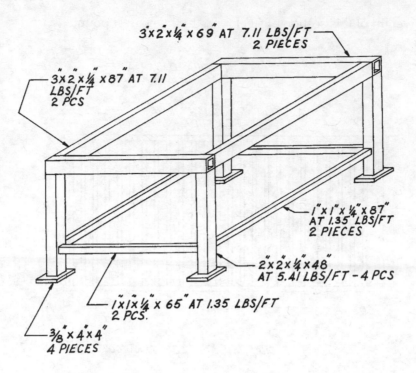

3″ x 2″ x ¼″ x 69″ AT 7.11 LBS/FT
2 PIECES

3″ x 2″ x ¼″ x 87″ AT 7.11 LBS/FT
2 PCS

1″ x 1″ x ¼″ x 87″ AT 1.35 LBS/FT
2 PIECES

2″ x 2″ x ¼″ x 48″ AT 5.41 LBS/FT – 4 PCS

1″ x 1″ x ¼″ x 65″ AT 1.35 LBS/FT
2 PCS.

⅜″ x 4″ x 4″
4 PIECES

Copyright by The Goodheart-Willcox Co., Inc.

Unit 22
Bending Metal

Key Terms

flat-out   outside corner

inside corner

Introduction

You will often be called upon to bend metal as part of your job in a welding shop. When a piece of metal is permanently bent, its original **flat-out** length is changed. Because of this dimensional change, you will have to be able to calculate flat-out lengths of metal objects that have been formed by bending. Most of the bends you will encounter can be grouped into the following types:

- 90° Sharp Corner Bend
- Circle
- Half Circle
- Quarter Circle

A short explanation of the effects of bending may help you understand the calculations of bending. Why does permanent bending affect the length of a piece of metal? The primary reason is because metal can be stretched and/or compressed. For example, when metal is bent, the thickness of the metal at the bend is compressed on the inside of the corner and stretched on the outside of the corner. The bent piece of metal now has two different lengths; the inside length and the outside length. The required accuracy of the part being produced will determine how accurately these lengths are to be calculated.

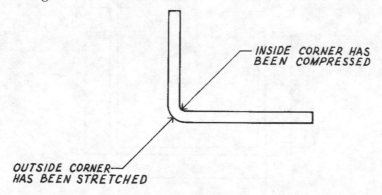

The ideas of inside corner and outside corner can be explained by considering perspective. In the following illustration, Observer A is looking at the outside corner of the bend. Observer B is looking at the inside corner of the bend. A corner is formed when two lines meet to form two angles: one on each side of the corner. From the side of an **inside corner**, the lines stretch forth at a less than 180° angle. Inside corners include, while outside corners exclude. Imagining the bend below as the walls of a house, it would appear that Observer B is inside the house, while Observer A is outside. From the side of an **outside corner**, the lines stretch forth at a greater than 180° angle.

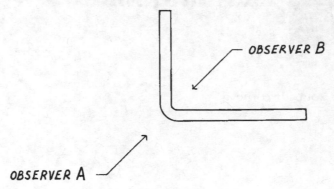

Bending metal can be a complicated manufacturing process involving specialized equipment and formulas. Welders, however, can use a few simple rules to quickly produce excellent, accurately bent parts from flat stock.

90° Sharp Corner Bend

By reshaping a single piece of metal, common shapes like squares and rectangles can be formed to include 90° sharp corner bends. It should be pointed out that all bends have a bend radius. Even the so-called 90° sharp corner bend has a bend radius although it is quite small. Because the radius is extremely small, the following rules of thumb assume a bend radius of zero.

Outside Dimensions

To calculate the flat stock length when given the outside dimensions of the part, subtract <u>twice</u> the metal thickness for each 90° bend. Notice the measurements in the diagram below are taken on the side of the outside corner of each bend.

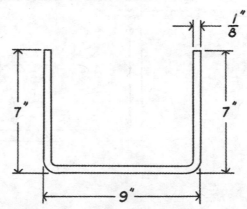

Subtract an amount equal to the metal thickness from the lengths of metal measured on each side of the bend.

Left	Center		Right
7″	9″	8⅞″	7″
− ⅛″	− ⅛″	− ⅛″	− ⅛″
6⅞″	8⅞″	8¾″	6⅞″

Flat length = 6⅞″ + 8¾″ + 6⅞″ = 22½″

Copyright by The Goodheart-Willcox Co., Inc.

The stretched out length is 22½". The flat layout would be as follows:

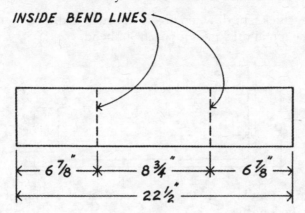

Inside Dimensions

To calculate the flat stock length when given the inside dimensions of the part, simply add the inside dimensions. This will give you the flat stock length. Notice the measurements in the diagram below are taken on the side of the inside corner of each bend.

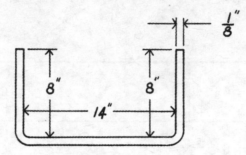

Flat length = 8" + 14" + 8" = 30"

The stretched out length is 30". The flat layout would be as follows:

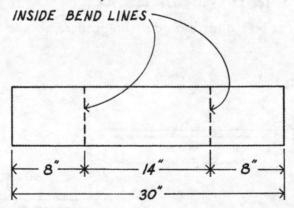

Copyright by The Goodheart-Willcox Co., Inc.

Outside and Inside Dimensions

To calculate the flat stock length when given one outside dimension and one inside dimension, subtract <u>one</u> metal thickness for each 90° bend.

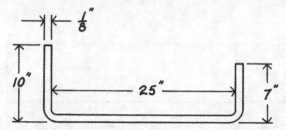

Since the piece in this diagram has two 90° bends, subtract the value of the metal thickness (⅛″) twice from the sum of the lengths.

$$\text{Flat length} = 10'' - \tfrac{1}{8}'' + 25'' + 7'' - \tfrac{1}{8}'' = 41\tfrac{3}{4}''$$

The stretched out length is 41¾″. The flat layout would be as follows:

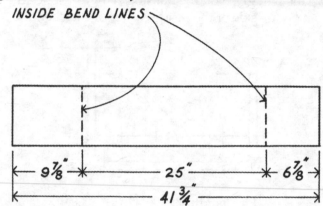

Review of 90° Sharp Corner Bends

The following part has been dimensioned in three different ways. The flat-out length is calculated as shown.

Inside Dimensions

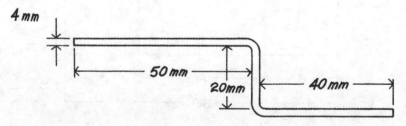

$$\text{Flat length} = 50 \text{ mm} + 20 \text{ mm} + 40 \text{ mm} = 110 \text{ mm}$$

Notice that only the inside corner length measurements were added to determine the flat-out length.

Copyright by The Goodheart-Willcox Co., Inc.

Outside Dimensions

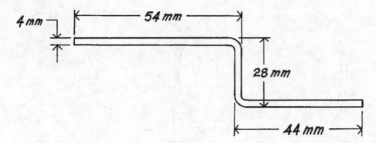

Flat length = 54 mm + 28 mm + 44 mm – 8 mm – 8 mm = 110 mm

Notice that the outside corner length measurements were added. Also, double the metal thickness was subtracted for each of the two 90° bends to determine the flat-out length.

Outside and Inside Dimensions

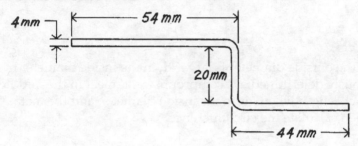

Flat length = 54 mm + 20 mm + 44 mm – 4 mm – 4 mm = 110 mm

Notice that two outside corner length measurements and one inside corner length measurement were added. Also, the metal thickness was subtracted once for each 90° bend to determine the flat-out length.

Circles

A piece of metal bent into a circle has two circumferences: inner and outer. The inside surface has been compressed into a smaller circumference, and the outside has been stretched into a larger circumference.

Outside Diameter

The difference between the outside and inside diameter of a metal piece bent into a circle is not the measurement of the metal thickness. It is twice the metal thickness. This is because the outside diameter crosses the metal thickness twice: once on each side of the measurement. Therefore, to calculate the flat stock length, we must reconcile the difference by meeting in the middle. When given the outside diameter, subtract the metal thickness once from the diameter and then calculate the circumference.

Copyright by The Goodheart-Willcox Co., Inc.

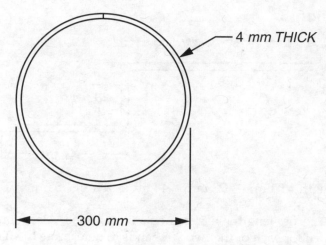

$$D - thickness = 300 \text{ mm} - 4 \text{ mm} = 296 \text{ mm}$$
$$\pi \times 296 \text{ mm} = 3.14 \times 296 \text{ mm} = 929 \text{ mm (rounded)}$$

Inside Diameter

Remember that the outside diameter is larger than the inside diameter by twice the metal thickness. As previously determined, we compromise and meet in the middle. Therefore, to calculate the flat stock length when given the inside diameter, add the metal thickness once to the diameter and then calculate the circumference.

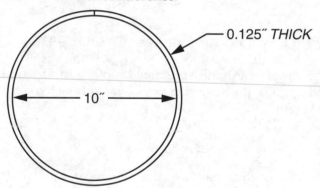

$$D + thickness = 10'' + 0.125'' = 10.125''$$
$$\pi \times 10.125'' = 3.14 \times 10.125'' = 31.7925'' = 31^{51}/_{64}'' \text{ (rounded)}$$

Half Circle

Half circle bends are calculated in the same way as full circle bends. The only difference is that one-half a circumference is calculated rather than a full circumference.

Outside Diameter

To calculate the flat stock length when given the outside diameter, subtract the thickness from the diameter, calculate the circumference, and then divide it in half.

Copyright by The Goodheart-Willcox Co., Inc.

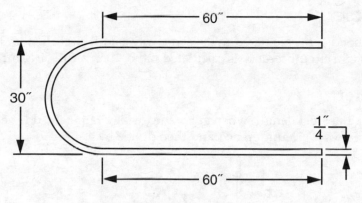

$$D - \text{thickness} = 30'' - \tfrac{1}{4}'' = 29\tfrac{3}{4}''$$

$$\text{Half Circumference} = \frac{\pi \times 29\tfrac{3}{4}''}{2} = \frac{3.14 \times 29.75''}{2} = 46\tfrac{45}{64}'' \ (\text{rounded})$$

$$\text{Flat length} = 60'' + 46\tfrac{45}{64}'' + 60'' = 166\tfrac{45}{64}'' \ (\text{rounded})$$

Inside Diameter

To calculate the flat stock length when given the inside diameter, add the thickness to the diameter, calculate the circumference, and then divide it in half.

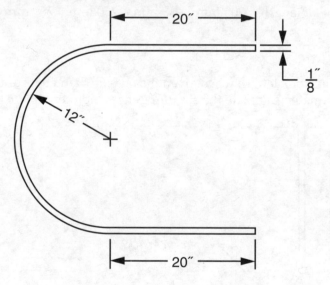

$$D + \text{thickness} = 24'' + \tfrac{1}{8}'' = 24\tfrac{1}{8}''$$

$$\text{Half Circumference} = \frac{\pi \times 24\tfrac{1}{8}''}{2} = \frac{3.14 \times 24.125''}{2} = 37\tfrac{7}{8}'' \ (\text{rounded})$$

$$\text{Flat length} = 20'' + 37\tfrac{7}{8}'' + 20'' = 77\tfrac{7}{8}''$$

Copyright by The Goodheart-Willcox Co., Inc.

Quarter Circle

Quarter circle bends are calculated in the same way as full circle bends. The only difference is that one-quarter of a circumference is calculated rather than a full circumference.

Inside Diameter

To calculate the flat stock length when given the inside diameter, add the metal thickness to the diameter, calculate the circumference, and then divide by 4.

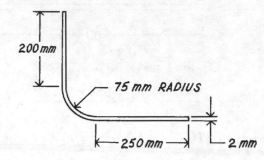

$$D + \text{thickness} = 150 \text{ mm} + 2 \text{ mm} = 152 \text{ mm}$$

$$\text{Quarter Circumference} = \frac{\pi \times 152 \text{ mm}}{4} = \frac{3.14 \times 152 \text{ mm}}{4} = 119 \text{ mm}$$

$$\text{Flat length} = 200 \text{ mm} + 119 \text{ mm} + 250 \text{ mm} = 569 \text{ mm}$$

Outside Diameter

To calculate the flat stock length when given the outside diameter, subtract the metal thickness from the diameter, calculate the circumference, and then divide by 4.

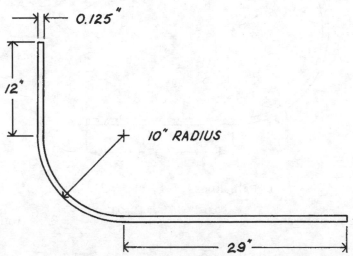

$$D - \text{thickness} = 20'' - 0.125'' = 19.875''$$

$$\text{Quarter Circumference} = \frac{\pi \times 19.875''}{4} = \frac{3.14 \times 19.875''}{4} = 15.6'' \text{ (rounded)}$$

$$\text{Flat length} = 12'' + 15.6'' + 29'' = 56.6''$$

Copyright by The Goodheart-Willcox Co., Inc.

Name _____ Date _____ Class _____

Unit 22 Practice

Perform the following equations as specified. Show all your work. Box your answers.

1. Seventeen brackets are required as shown. What is the total length of metal required?

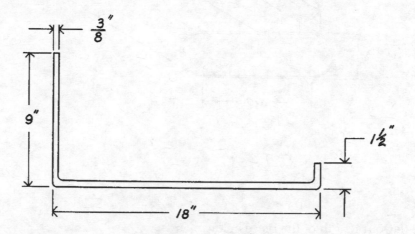

2. Calculate the flat-out dimensions (length and width) of part (A).

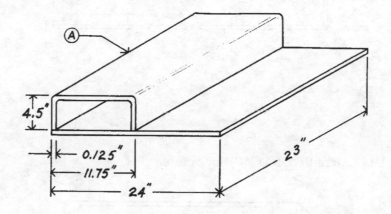

3. Calculate the flat-out length.

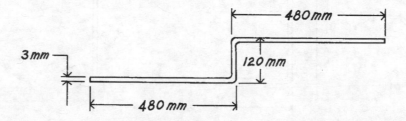

Copyright by The Goodheart-Willcox Co., Inc.

4. What is the flat-out length and width of this ¼″ thick tank to the nearest thirty-second?

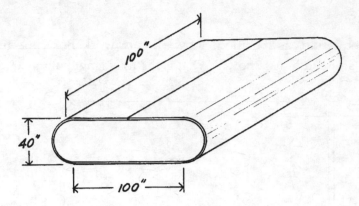

5. Calculate the flat-out length of the piece below.

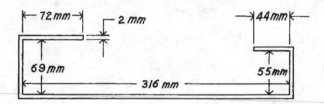

6. Calculate the flat-out length of this ³⁄₁₆″ thick piece.

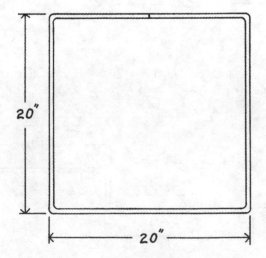

Copyright by The Goodheart-Willcox Co., Inc.

Name _____ Date _____ Class _____

7. Calculate the flat-out length of this ³⁄₁₆″ thick piece.

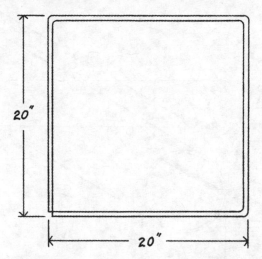

8. Calculate the flat-out length of this 3 mm thick piece.

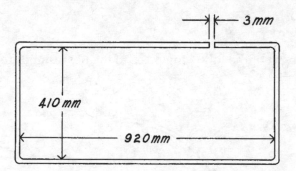

9. Calculate the flat-out length of this ¼″ thick ring to the nearest thirty-second.

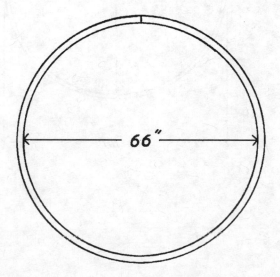

Copyright by The Goodheart-Willcox Co., Inc.

10. Calculate the flat-out dimension for this piece to the nearest sixty-fourth.

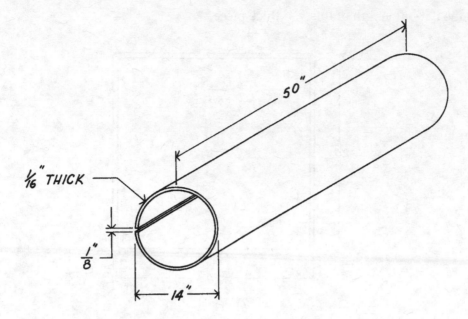

11. Calculate the flat-out length of the following piece to the nearest millimeter.

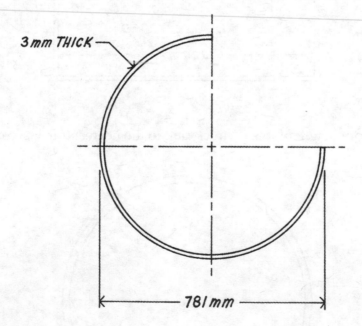

Copyright by The Goodheart-Willcox Co., Inc.

Name _____ Date _____ Class _____

12. Calculate the flat-out length of this 2 mm thick piece of steel.

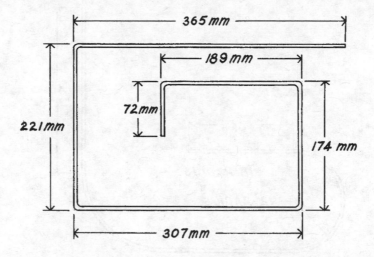

13. Calculate the flat-out length. The metal is ¹⁄₁₆″ thick.

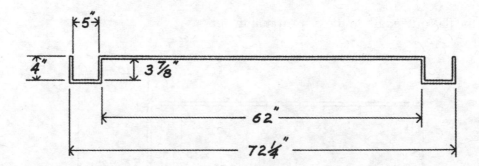

Copyright by The Goodheart-Willcox Co., Inc.

14. Calculate the flat-out length to the nearest sixty-fourth.

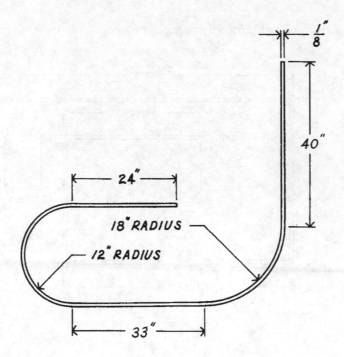

15. Calculate the flat-out length to the nearest millimeter.

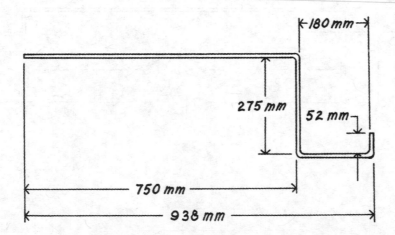

Copyright by The Goodheart-Willcox Co., Inc.

Section 6
Percentages and the Metric System

Section Objectives

After studying this section, you will be able to:

- Define percentage
- Convert between a fraction, a decimal, and a percentage
- Calculate percentages
- List the seven basic metric units
- Explain the derived units of measure from length and mass
- List the standard units of metric length
- Convert measurements of length
- Estimate and visualize metric lengths
- Explain derived units of metric length
- List the standard units of metric mass
- Convert measurements of mass
- Estimate and visualize metric mass
- Explain derived units of metric mass
- Demonstrate the proper way to use the Language of SI

Copyright by The Goodheart-Willcox Co., Inc.

Unit 23
Percentages

Key Term

percent (%)

Introduction

Percent is a word that expresses things "by the hundred." Percents are parts of the whole of anything when it is divided into 100 equal parts. For example, 80 parts out of 100 can be expressed as 80%. The % symbol next to a number tells you it is a percentage number. The use of percents is very common in daily working life. Items such as the following are usually expressed in percentages:

- The composition of steel
- The composition of alloys
- Salary deductions
- Scrap rate on a job
- Quality control statistics

Being able to solve problems involving percentage is a useful skill. Mathematically, a variety of things can be done with percent. Percents can be converted to decimals or fractions and vice versa. The percent of a number can be calculated. One number can be expressed as a percent of another.

Changing a Percent to a Fraction

Since percents refer to hundredths, changing them to fractions is quite easy. Simply remove the percent sign and write the number as a fraction with a denominator of 100. Reduce if necessary.

$$75\% = \frac{75}{100} = \frac{3}{4}$$

$$100\% = \frac{100}{100} = 1$$

$$125\% = \frac{125}{100} = 1\frac{25}{100} = 1\frac{1}{4}$$

$$3,000\% = \frac{3,000}{100} = 30$$

The above method works fine for most cases, but if the percent is a fraction or a mixed number, there is an additional step. First, change the mixed number to an improper fraction. Now that you have the percent in fractional form, remove the percent sign, multiply the denominator by 100, and then reduce, if necessary.

$$\frac{1}{4}\% = \frac{1}{400}$$

$$\frac{3}{16}\% = \frac{3}{1600}$$

Copyright by The Goodheart-Willcox Co., Inc.

$$12\frac{1}{2}\% = \frac{25}{2}\% = \frac{25}{200} = \frac{1}{8}$$

$$107\frac{1}{2}\% = \frac{215}{2}\% = \frac{215}{200} = 1\frac{15}{200} = 1\frac{3}{40}$$

Changing a Percent to a Decimal

This is done using the same method as explained for changing percents to fractions, except the final answer is expressed as a decimal number. Remove the percent sign, divide the number by 100, and then express the answer as a decimal.

$$67\% = \frac{67}{100} = 0.67$$

$$16.6\% = \frac{16.6}{100} = 0.166$$

$$219\% = \frac{219}{100} = 2.19$$

$$1{,}019\% = \frac{1{,}019}{100} = 10.19$$

In cases where the percent is a fraction or a mixed number, first change the fraction to a decimal.

$$\frac{1}{16}\% = 0.0625\% = \frac{0.0625}{100} = 0.000625$$

$$11\frac{1}{8}\% = 11.125\% = \frac{11.125}{100} = 0.11125$$

$$95\frac{1}{2}\% = 95.5\% = \frac{95.5}{100} = 0.955$$

$$365\frac{1}{4}\% = 365.25\% = \frac{365.25}{100} = 3.6525$$

Changing a Fraction to a Percent

The easiest method of changing fractions, mixed numbers, and whole numbers to percents is to multiply by 100 and add a percent sign.

$$\frac{9}{10} = \frac{9}{10} \times 100 = 90\%$$

$$1 = 1 \times 100 = 100\%$$

$$2\frac{3}{4} = \frac{11}{4} \times 100 = 275\%$$

$$5\frac{3}{7} = \frac{38}{7} \times 100 = 542\frac{6}{7}\%$$

Copyright by The Goodheart-Willcox Co., Inc.

Changing a Decimal to a Percent

To change decimals to percents, multiply by 100 and add a percent sign.

$$0.01 = 0.01 \times 100 = 1\%$$

$$0.25 = 0.25 \times 100 = 25\%$$

$$1.0 = 1.0 \times 100 = 100\%$$

$$3.5 = 3.5 \times 100 = 350\%$$

Calculating a Percent of a Number

This type of question is usually expressed as "find 25% of 300." You may recall from Unit 9, *Multiplication of Fractions*, that "of" means to multiply. Therefore, the solution is arrived at by multiplying 25% × the number in question. In this case, it is 300. To do this, change the percent to a decimal and proceed with the multiplication. In this example, $0.25 \times 300 = 75$. Twenty-five percent of 300 is 75.

The percent can also be calculated in fractional form rather than as a decimal. This is especially useful if the decimal form is a repeating decimal such as 0.666 . . .

$$30\% \text{ of } 3{,}750 = 0.30 \times 3{,}750 = 1{,}125$$

$$112\% \text{ of } 43 = 1.12 \times 43 = 48.16$$

$$1.5\% \text{ of } 900 = 0.015 \times 900 = 13.5$$

$$16\frac{2}{3}\% \text{ of } 2{,}700 = \frac{50}{3}\% \text{ of } 2{,}700 = \frac{50}{300} \times 2{,}700 = 450$$

Calculating the Percentage One Number Is of Another Number

This type of question is usually expressed as, "What percent of 52 is 13?" The word "of" means "to multiply," and "is" means "equals." Therefore, the questions can be expressed in the following way.

$$?\% \times 52 = 13$$

To solve this problem, isolate the unknown percent value. This can be achieved by dividing both sides of the equation by 52.

$$\frac{?\% \times 52}{52} = \frac{13}{52}$$

On the left side of the equal symbol, the 52 below and the 52 above canceled out each other.

$$\frac{?\% \times \overset{1}{\cancel{52}}}{\underset{1}{\cancel{52}}} = \frac{13}{52}$$

This will form a fraction with the multiplier (52) as the denominator and 13 as the numerator. Since we are looking for a percent, we know that the denominator under the unknown value will be 100.

$$\frac{?}{100} = \frac{13}{52}$$

Copyright by The Goodheart-Willcox Co., Inc.

This equation can then be cross-multiplied.

$$\frac{?}{100} \quad \frac{13}{52}$$

We now need only to divide and plug the result back into the original statement: ?% of 52 is 13. Remember that "of" means "to multiply" and "is" means "equals."

$$\frac{1,300}{52} = 25$$

$$25\% \times 52 = 13$$

$$25\% \text{ of } 52 \text{ is } 13$$

The way in which this type of question is expressed may cause some confusion. For example, the question was originally expressed as "What percent of 52 is 13?" It could also have been expressed as, "Thirteen is what percent of 52?" Since "is" means equals and "of" means times, the resulting equation is $13 = ?\% \times 52$. Both expressions have the same meaning.

Copyright by The Goodheart-Willcox Co., Inc.

Work Space

Copyright by The Goodheart-Willcox Co., Inc.

Name _____ Date _____ Class _____

Unit 23 Practice

Express the following percents as fractions. Show all your work. Be certain the columns line up. Box your answers.

1. 13%

2. 0.3%

3. 0.05%

4. $\dfrac{1}{16}$ %

5. 1010%

6. $\dfrac{2}{7}$ %

Express the following percents as decimals. Show all your work. Be certain the columns line up. Box your answers.

7. 75%

8. 39%

9. 14.14%

10. 0.125%

11. 0.007%

12. $\dfrac{2}{5}$ %

Express the following fractions as percents. Show all your work. Be certain the columns line up. Box your answers.

13. ¾

14. ¹⁄₁₀₀

15. 33⅓

16. 2230½

17. ¹¹⁄₁₉

18. 99¹⁄₁₆

Copyright by The Goodheart-Willcox Co., Inc.

Express the following decimals as percents. Show all your work. Be certain the columns line up. Box your answers.

19. 0.125   20. 1.125

21. 0.019   22. 140.7

23. 0.003   24. 0.89

Calculate the following to two decimal places. Show all your work. Be certain the columns line up. Box your answers.

25. 35% of 823   26. 9.4% of 507

27. 100% of 28.5   28. $33\frac{1}{3}$% of 253

29. 0.17% of 123.6   30. 1973% of 22

31. What percent of 423 is 95?   32. 579 is what percent of 606?

33. One (1) is what percent of 2?   34. What percent of 15 is 138?

35. What percent of 50 is 450?   36. 31.2 is what percent of 65.7?

37. A welder's suggestion to use a different type of fixture resulted in a 15% increase in production. The previous production rate was 180 parts per hour. What is the new production rate?

38. For year one, a group of mills produced 83 million tons of steel. For the second year, production increased by 8.5%. How many tons of steel were produced in the second year?

Copyright by The Goodheart-Willcox Co., Inc.

Name _____ Date _____ Class _____

39. Three hundred and eighteen motorcycles started in the Illinois-Michigan Motorcycle Rally. Only eighty-five bikes finished the rally. What percent of the bikes did not finish? Calculate to the nearest percent.

40. A welding shop was allowed a discount of 12 1/2% on the regular price of $426.00 for a pedestal grinder. Calculate the price the shop paid to the nearest cent.

41. The website for the Metal Building Association reported that sales this year were 32 1/2% ahead of last year's total of $1,266,000,000. What was the dollar value increase in metal building sales this year?

Use the following information to calculate totals to fill in the table below. All of these employees listed received a salary increase of 4.75%. Calculate the weekly gross pay, income tax, union dues, and social security to the nearest cent.

	Employees	Weekly Gross Pay		Income Tax		Union Dues	Social Security
		Before increases	After increases	%	$	3½% of gross pay	4% of gross pay
42.	Calleja, Nick	$453.60		16%			
43.	Ing, Richard	$491.60		18%			
44.	Lima, Jose	$354.00		15%			
45.	Massey, Vincent	$320.00		15%			
46.	Shulman, Lorna	$540.80		19%			
47.	Simpson, Katie	$402.00		18%			
48.	Williams, Deroy	$458.80		15%			
49.	Zakoor, Eli	$398.80		17%			

Copyright by The Goodheart-Willcox Co., Inc.

Work Space

Copyright by The Goodheart-Willcox Co., Inc.

Unit 24
The Metric System

A Metric Workplace

Anyone who has been involved with the metal working trades realizes that the global market is a powerful force. The decision by major manufacturers to employ the metric system has guaranteed its prominence in the workplace.

The welding industry is just one of many industries to adopt the metric system. As a welder, you will be expected to understand and work in metric. You will be pleased to know it is not a difficult system to learn. It is a very logical, very consistent system. However, the biggest difficulty most people have is changing established habits. Since many of us are so accustomed to using the inch/lb units, we find that giving them up can be quite difficult. But, as job requirements include "the ability to work in metric," an attachment to the inch/lb system usually gives way to the reality that an understanding of metrics is an important part of your training as a welder.

The Metric System

The metric system, since it is a measuring system, is used for all types of measurement. In fact, the metric system (also referred to as SI; see the Glossary) is intended to be a system of measurement for everything measurable. The creation of SI was a worldwide cooperative undertaking. Committees of experts were formed to find the best possible way of measuring. All this effort resulted in the establishment of what is known as the seven basic units of SI. The outstanding characteristic of the seven basic units is that any physical quantity in the universe can be measured by one of the seven basic units.

The Basic Units

The seven basic units of the metric system are as follows:

- Meter (m)—measuring length
- Kilogram (kg)—measuring mass
- Second (s)—measuring time
- Ampere (A)—measuring electrical current
- Kelvin (K)—measuring temperature
- Candela (cd)—measuring luminous intensity
- Mole (mol)—measuring the amount of substance

In welding, you will be required to know and apply the two basic units of length and mass. In everyday use, the word *weight* is often used in place of the word *mass*, although they have different meanings (see the Glossary). Be aware of the differences. However, in this text, we will use the word *weight* as it is often understood in everyday speech, having the same meaning of the term *mass*.

Derived Units of Measure

The basic units are subdivided into parts called derived units. These derived units each measure different dimensions. For example, the dimension of area is a function of two different dimension of length. The two units of length (meter) combine (meter × meter) to form a new,

Copyright by The Goodheart-Willcox Co., Inc.

derived unit of length (meter^2). Review these derived dimensions and units of length and mass below.

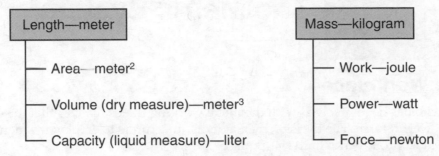

Length—meter	Mass—kilogram
— Area—meter^2	— Work—joule
— Volume (dry measure)—meter^3	— Power—watt
— Capacity (liquid measure)—liter	— Force—newton

It is usually not necessary for a welder to be completely familiar with the derived units of mass. However, the derived units of length are important to know.

Length

Length is one of the basic dimensions you will work with in the metric world. The originators of SI decided each unit of measure should have a base unit as a reference point. Then, subdivisions and multiples of the base unit were agreed upon and named. Their lengths and names were all related to the base unit, making a very consistent system. The meter was chosen as the base unit for length.

Standardized Units of Length

The following chart shows the base unit (the meter) and some of the subdivisions and multiples established. The very large multiples and the very small subdivisions have been omitted from the chart in order to focus on those most common in ordinary daily life.

Name	Symbol	How Each Unit of Length Is Related to the Base Unit (Meter)
Kilometer	km	1000 meters
Hectometer	hm	100 meters
Dekameter	dam	10 meters
Meter	m	1 meter
Decimeter	dm	0.1 meter
Centimeter	cm	0.01 meter
Millimeter	mm	0.001 meter

Base unit →→ (points to Meter row)

Start reading the chart at the base unit (the meter) and work your way up and down from there. Do not be concerned about what these lengths actually look like. You will get familiar with this later. For now, just observe how well the system is organized. The names of all units are related by their suffixes. They all end in *meter*. Even the abbreviations are related; they all end in the letter *m*. Also, all units are related to the base unit in multiples of 10. In welding, you will not be working with the hectometer, dekameter, and decimeter. They are illustrated here so that you can see the regularity and pattern of the system.

Copyright by The Goodheart-Willcox Co., Inc.

Converting Metric Units of Length

The following diagram is a useful device for learning how to convert metric measurements of length.

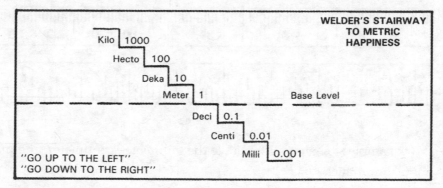

1. Convert 1629 m to kilometers.
   Looking at the stairway, you have to go up three steps (or places) to the left to move from m to kilometers. So, in this case, you would move the decimal point three places to the left. Therefore, 1629 m equals 1.629 kilometers.

2. Convert 85 cm to millimeters.
   Looking at the stairway, you have to go down one step to the right to move from cm to millimeters. In this case, you would move the decimal one place to the right. Therefore, 85 cm equals 850 mm.

3. Convert 7.5 km to decimeters.
   Go down the stairway four places to the right. Move the decimal four places to the right. Therefore, 7.5 km equals 75000 dm.
   The "metric stairway" is a useful visual tool for converting measurements. There is, of course, a mathematical explanation why conversion from one unit to another can be done simply by moving the decimal. Since all units are related in multiples of ten, converting consists of multiplying by these multiples and, in effect, moving the decimal.

4. Convert 5.9 km to meters.
   5.9 km × 1000 = 5900 m

5. Convert 16 cm to millimeters.
   16 cm × 10 = 160 mm

6. Convert 1437 dm to hectometers.
   1437 dm × 0.001 = 1.437 hm

Visualizing Metric Lengths

Becoming knowledgeable in the metric system requires that you develop an awareness of the actual lengths of metric units. Just as you already have a rough idea of the length of one inch, one foot, and one yard, you can also start becoming aware of metric lengths. In order to avoid being overwhelmed with all the units of length, it is best to concentrate just on the millimeter, centimeter, and meter. Being familiar with these three will satisfy your needs in

Copyright by The Goodheart-Willcox Co., Inc.

welding. In fact, almost all the metric blueprints you will be reading are dimensioned only in millimeters. Below is a full size metric scale 150 mm long. If you do not own a metric scale, you should buy one now.

(Herlihy)

Here are a few examples to help you visualize the millimeter, centimeter, and meter.

- A dime is about 1 mm thick.
- A thick line drawn with a dull pencil is about 1 mm wide.
- A slice of bread is about 1 cm thick.
- A stack of ten dimes is about 1 cm high.
- One meter is about three inches longer than one yard.
- The height of a doorknob is about one meter from the floor.

Estimating Metric Lengths

A useful technique in becoming knowledgeable in the metric system is to memorize certain body dimensions, then use these to estimate other dimensions.

Measure the following to the nearest millimeter (mm) and record the results.

> The width of your thumb in millimeters:
>
> The width of your palm in millimeters:
>
> The span of your hand from the smallest finger to thumb in centimeters:
>
> The distance from your elbow to your fingertips in centimeters:
>
> The length of your stride in centimeters:
>
> Your height in meters:

Now try the estimating exercise below.

1. Estimate, using the span of your hand as a guide, the length of an object and the number of times it takes to span the length with your hand.

2. Multiply the number of spans by the length of a span to arrive at an estimate of the length. If your hand has a span of 24 cm and an object is 5 spans long, its estimated length is 24 times 5 or 120 cm.

3. Check your accuracy with a metric scale.

Converting Metric and US Customary Units of Length

At some time in your job, you will probably have to make conversions between these two systems. Conversion charts or cards are often available, but you should know the mathematical method for converting. Your main concern as a welder will be converting from mm to inches and from inches to mm. Therefore, these are the only conversions that will be explained here.

Copyright by The Goodheart-Willcox Co., Inc.

Calculations in converting often lead to the necessity of rounding the answer. Answers can be rounded for maximum mathematical accuracy or they can be rounded for the accuracy normally required in the shop. A method for shop accurate conversions will be explained here.

Shop Acceptable Accuracy

In general, it can be stated that the closest tolerance to which a welder may be expected to work is to the nearest $\frac{1}{64}$ inch. The following guidelines ensure that your work stays within this degree of accuracy.

1. When converting from inches to mm, round off to the nearest mm.

2. When converting from mm to inches, round off to the nearest 0.1". Below are some examples:
    Convert 7.367" to mm
    Exact conversion: $7.367 \times 25.4 = 187.1218$ mm
    Shop acceptable accuracy = 187 mm

    Convert $11\frac{1}{16}$" to mm
    Exact conversion: $11.0625 \times 25.4 = 280.9875$ mm
    Shop acceptable accuracy = 281 mm

    Convert 65.5 mm to inches
    Exact conversion: $65.5 \div 25.4 = 2.5787402$"
    Shop acceptable accuracy = 2.6"

    Convert 485.35 mm to inches
    Exact conversion: $485.35 \div 25.4 = 19.108268$"
    Shop acceptable accuracy = 19.1"

Two rules for converting are stated below:

*   When converting from inches to millimeters, multiply by 25.4.

*   When converting from millimeters to inches, divide by 25.4.

Derived Units of Length

As mentioned earlier, the basic unit of length has three derived units of length: area, volume, and capacity.

Area

Area is calculated in the normal manner, depending upon the shape of the object. Results are expressed as mm^2, km^2, etc. Be sure you are calculating with the same units; that is, mm $\times$ mm or cm $\times$ cm. For example, do not multiply meters times millimeters to calculate an area.

Volume

Volume refers to dry measure and is calculated in the normal manner, depending upon the shape of the object. Results are expressed as m^3, mm^3, etc. As pointed out with area, be certain you are calculating with the same units: m $\times$ m $\times$ m = m^3.

Copyright by The Goodheart-Willcox Co., Inc.

Capacity

Capacity is a derived unit of length used for liquid measure. The base unit is the liter. The standardized units of capacity are illustrated in the chart below.

Name	Symbol	How Each Unit of Capacity Is Related to the Base Unit (Liter)
Kiloliter	kl	1000 liters
Hectoliter	hl	100 liters
Dekaliter	dal	10 liters
Liter	L	1 liter
Deciliter	dl	0.1 liter
Centiliter	cl	0.01 liter
Milliliter	ml	0.001 liter

Base unit ⟶ (Liter)

As you saw with the chart for units of length, there is a regular pattern of units of capacity. Also, as with units of length, you will work with only a few of the units of capacity, such as the liter (L) and milliliter (ml). The other units are illustrated here so you can see the pattern of the system.

Converting Metric Units of Capacity

To convert from one unit of measure to another, use the following diagram. Also, see the examples below.

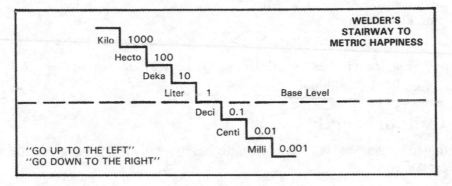

Convert 49 liters (L) to milliliters (ml)
49 L = 49000 ml

Convert 7 kiloliters (kl) to liters
7 kl = 7000 L

Convert 28 milliliters to liters
28 ml = 0.028 L

Visualizing Metric Capacity

Here are a few examples to help you visualize the liter and milliliter.
- A teaspoon is about 5 ml
- A bowl of soup is about 200 ml

Copyright by The Goodheart-Willcox Co., Inc.

- The liter is approximately the volume of 4 paper coffee cups
- A tea kettle has a capacity of about 2 L

Converting Metric and US Customary Units of Capacity

To convert between metric and standard capacity, use the conversion figure of one gallon = 3.785 liters.

Convert 27 gallons to liters (L)
27 gallons = 102.195 L

Convert 16½ gallons to liters (L)
16½ gallons = 62.4525 L

Two rules for converting between metric and US Customary units of capacity:

- When converting from gallons to liters, multiply by 3.785.
- When converting from liters to gallons, divide by 3.785.

Mass

Mass is one of the basic units you will work with in the metric system. The base unit for mass is the gram. The following chart shows the base unit and some of the subdivisions and multiples established for metric mass. The very large multiples and the very small subdivisions have been omitted in order to focus on those most common in ordinary daily life.

Name	Symbol	How Each Unit of Mass Is Related to the Base Unit (Gram)
Megagram	Mg (t)	1000000 grams
—	—	—
—	—	—
Kilogram	kg	1000 grams
Hectogram	hg	100 grams
Dekagram	dag	10 grams
Gram *(Base unit →)*	g	1 gram
Decigram	dg	0.1 gram
Centigram	cg	0.01 gram
Milligram	mg	0.001 gram

Start reading the chart at the base unit and work your way up and down from there. Observe how well the system is organized. All units are related to the base unit in multiples of 10. The names of all units are related by their suffixes; they all end in *gram*. Even the abbreviations are related; they all end in the letter *g*. There is one variation accepted here. The megagram is also commonly referred to as a metric ton and has the symbol *t*.

Of the eight units of mass shown on the chart, you will probably encounter only the megagram (metric ton), kilogram, gram, and milligram. The other units are illustrated here so that you can see the pattern of the system. Since the units increase in multiples of 10, you may have expected a metric unit for 10000 grams and 100000 grams. However, there are no metric units between the kilogram (1000 grams) and the metric ton (1000000 grams).

Copyright by The Goodheart-Willcox Co., Inc.

Converting Metric Units of Mass

The following diagram is a useful device for learning how to convert measurements. Review it before reading through the examples below.

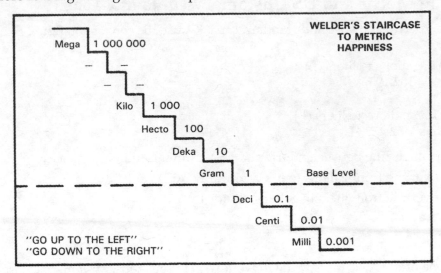

Convert 3275 g to kilograms (kg)
You have to go up three steps (or places) to the left to move from gram to kilograms. So, you would move the decimal point three places to the left. Therefore, 3275 g equals 3.275 kg.

Convert 987000 milligrams (mg) to kilograms
You have to go up six places to the left to move from milligrams to kilograms. Move the decimal six places to the left. Therefore, 987000 mg equals 0.987 kg.

Convert 48 megagrams (Mg) to kilograms
You have to move down three places to the right to move from megagrams to kilograms. Move the decimal three places to the right. Therefore, 48 Mg equals 48000 kg.

As you can see from these and previous examples, converting between units in metric can be done simply by moving the decimal point.

Visualizing Metric Mass

The metric weights you should become familiar with are gram, kilogram, and metric ton. Below are some examples:

- A paper clip has a mass of about 1 gram
- A thumbtack weighs about 1 gram
- The mass of about five average sized apples is about 1 kilogram
- A heavyweight boxer weighs about 91 kilograms
- A compact car has a mass of about 1 metric ton
- One metric ton weighs about 200 lb more than 1 standard ton

Converting Metric and US Customary Units of Mass

1 lb	= 0.4536 kg
1 kg	= 2.205 lb
1 metric ton (t)	= 2,205 lb

Copyright by The Goodheart-Willcox Co., Inc.

The above conversions have been rounded and, therefore, slight inaccuracies will result. They are accurate enough, however, for the work you will encounter in the shop. Below are some examples.

Convert 875 lb to kilograms
$875 \times 0.4536 = 396.9$ lb

Convert 32,000 lb to metric tons
$32,000 \div 2,205 = 14.5$ metric tons (rounded)

Convert 81.8 kg to lb
$81.8 \times 2.205 = 180.4$ lb (rounded)

Convert 100 000 kg to standard tons
$$\frac{100,000 \times 2.205}{2,000} = 110.25 \text{ tons}$$

Convert 5 metric tons to lb
$5 \times 2205 = 11,025$ lb

Convert 30 metric tons to standard tons
$$\frac{30 \times 2,205}{2,000} = 33 \text{ standard tons (rounded)}$$

The Language of SI

1. Never use a period after a symbol unless it is at the end of a sentence.
   Example: Her height is 148 cm, but his height is 185 cm.
2. Symbols are never made plural by adding an *s*.
   Example: 1 g   10 g   100 g   1000 g
3. Symbols are almost always written in lowercase. The two exceptions so far are megagram and liter.
   Example: m for meter, g for gram, but Mg for megagram and L for liter.
4. Never start a sentence with a symbol. Write the name out in full.
   Example: Kilometers measure distance.
5. Always leave a space between the quantity and the symbol.
   Example: 15 km   6 m   24 g
6. If there is no quantity with the unit, spell the unit out in full.
   Example: Meters are marvelous, and grams are great.
7. Always use decimals, not fractions.
   Example: 0.25 g   1.5   2.75 m
8. Always place a zero before the decimal point if the value is less than one.
   Example: 0.125 L   0.75 kg   0.5 m

Copyright by The Goodheart-Willcox Co., Inc.

Work Space

Copyright by The Goodheart-Willcox Co., Inc.

Name _____ Date _____ Class _____

Unit 24 Practice

1. List the seven basic units in the metric system.

2. Which two of the seven basic dimensions will welders encounter most often?

3. Name the base unit for metric length.

4. How many millimeters are there in one meter?

5. How many centimeters are there in one meter?

Convert the following measurements. Show all your work. Box your answers.

6. 3.5 kilometers to millimeters

7. 1597 centimeters to meters

8. 12800 millimeters to kilometers

9. 2985400 meters to kilometers

10. 1000 centimeters to millimeters

11. 595 decimeters to kilometers

Convert the following measurements to the nearest millimeter. Show all your work. Box your answers.

12. 0.375″ to millimeters

13. 36″ to millimeters

14. 5′-8½″ to millimeters

15. 16½″ to millimeters

16. 100′ to meters

17. 100 yd. to meters

Copyright by The Goodheart-Willcox Co., Inc.

Convert the following measurements to the nearest ¹⁄₆₄ inch. Show all your work. Box your answers.

18. 1000 mm to inches

19. 87.6 mm to inches

20. 112.8 mm to inches

21. 1000 m to feet and inches

22. 3855 mm to feet and inches

23. 10 km to feet and inches

24. Name the base unit for liquid capacity.

25. How many milliliters are there in a liter?

Convert the following. Show all your work. Box your answers.

26. 7570 L to gallons

27. 182 ml to liters

28. 54 gal. to liters

29. 6.7 L to milliliters

30. 1 gal. to milliliters

31. 10000 ml to gallons (to nearest hundredth)

32. Name the base unit for mass.

33. How many grams are there in one kilogram?

Convert the following. Show all your work. Box your answers.

34. 76950 g to milligrams

35. 39.25 kg to grams

36. 0.78 kg to milligrams

37. 212 mg to grams

38. 1000 mg to kilograms

39. 65 g to kilograms

Copyright by The Goodheart-Willcox Co., Inc.

Name _____ Date _____ Class _____

Convert the following. Show all your work. Box your answers.

40. 1,000 lb to kilograms

41. 6472 g to lb (round to nearest hundredth)

42. 105 lb to grams

43. 3.5 standard tons to kilograms (round to nearest kilogram)

44. 4.25 t to standard tons (round to nearest hundredth)

45. 69 standard tons to kilograms (round to nearest kilogram)

Copyright by The Goodheart-Willcox Co., Inc.

Work Space

Copyright by The Goodheart-Willcox Co., Inc.

Appendix

Useful Information

The following pages include several charts and tables filled with information that will prove useful to you throughout your career as a welder. This information includes decimal and fractional equivalents, conversions between US Customary and metric measurements, steps and calculations for bending metal, and a summary of formulas. Become familiar with this useful information.

Fraction, Decimal, and Millimeter Conversions					
Fractional Inch	**Decimal Inch**	**Millimeter**	**Fractional Inch**	**Decimal Inch**	**Millimeter**
1/64	.015625	0.397	33/64	.515625	13.097
1/32	.03125	0.794	17/32	.53125	13.494
3/64	.046875	1.191	35/64	.546875	13.891
1/16	.0625	1.588	9/16	.5625	14.288
5/64	.078125	1.984	37/64	.578125	14.684
3/32	.09375	2.381	19/32	.59375	15.081
7/64	.109375	2.778	39/64	.609375	15.478
1/8	.125	3.175	5/8	.625	15.875
9/64	.140625	3.572	41/64	.640625	16.272
5/32	.15625	3.969	21/32	.65625	16.669
11/64	.171875	4.366	43/64	.671875	17.066
3/16	.1875	4.762	11/16	.6875	17.462
13/64	.203125	5.159	45/64	.703125	17.859
7/32	.21875	5.556	23/32	.71875	18.256
15/64	.234375	5.953	47/64	.734375	18.653
1/4	.25	6.350	3/4	.75	19.05
17/64	.265625	6.747	49/64	.765625	19.447
9/32	.28125	7.144	25/32	.78125	19.844
19/64	.296875	7.541	51/64	.796875	20.241
5/16	.3125	7.938	13/16	.8125	20.638
21/64	.328125	8.334	53/64	.828125	21.034
11/32	.34375	8.731	27/32	.84375	21.431
23/64	.359375	9.128	55/64	.859375	21.828
3/8	.375	9.525	7/8	.875	22.225
25/64	.390625	9.922	57/64	.890625	22.622
13/32	.40625	10.319	29/32	.90625	23.019
27/64	.421875	10.716	59/64	.921875	23.416
7/16	.4375	11.112	15/16	.9375	23.812
29/64	.453125	11.509	61/64	.953125	24.209
15/32	.46875	11.906	31/32	.96875	24.606
31/64	.484375	12.303	63/64	.984375	25.003
1/2	.5	12.700	1	1.0	25.400

Copyright by The Goodheart-Willcox Co., Inc.

Useful Conversions between the SI Metric and US Customary Systems				
Length	1 inch	= 25.4 mm	1 mm	= 0.03937 inches
	1 inch	= 2.54 cm	1 cm	= 0.3937 inches
	1 foot	= 30.48 cm	1 cm	= 0.0328 feet
	1 foot	= 0.3048 m	1 m	= 3.28 feet
	1 yard	= 0.9144 m	1 m	= 1.0936 yard
	1 mile	= 1.609 km	1 km	= 0.621 mile
Area	1 in^2	= 645.2 mm^2	1 mm^2	= 0.00155 in^2
	1 in^2	= 6.452 cm^2	1 cm^2	= 0.155 in^2
	1 ft^2	= 929.03 cm^2	1 cm^2	= 0.00108 ft^2
	1 ft^2	= 0.093 m^2	1 m^2	= 10.764 ft^2
	1 yd^2	= 0.836 m^2	1 m^2	= 1.196 yd^2
	1 square mile	= 2.59 km^2	1 km^2	= 0.386 square mile
Volume (dry)	1 in^3	= 16.388 cm^3	1 cm^3	= 0.061 in^3
	1 ft^3	= 0.028 m^3	1 m^3	= 35.31 ft^3
	1 yd^3	= 0.765 m^3	1 m^3	= 1.308 yd^3
Capacity (fluid)		U.S.	Canadian	
	1 quart	= 0.946 L	1.14 L	
	1 gallon	= 3.785 L	4.546 L	
	1 L	= 1.057 quarts	0.877 quarts	
	1 L	= 0.264 gallons	0.220 gallons	
Weight	1 oz	= 28.35 g	1 g	= 0.0353 oz
	1 lb	= 453.59 g	1 kg	= 2.205 lb
	1 lb	= 0.4536 kg	1 ton (metric)	= 2,204.6 lb

Note: Converting between the US Customary and metric systems often results in highly unwieldy numbers. Because of this, many of the above conversions have been rounded, but are quite accurate as a quick reference.

Copyright by The Goodheart-Willcox Co., Inc.

Bending Metal

(How to Calculate Flat Stock Lengths Assuming a Zero Bend Radius)

Half Circle

1. Given the outside diameter, subtract the thickness from the diameter and then calculate one-half a circumference; or
2. Given the inside diameter, add the thickness to the diameter and then calculate one-half a circumference.

90° Corner Bend

1. Given the outside dimensions, add the outside dimensions and subtract twice the metal thickness for each 90° bend; or
2. Given the inside dimensions, add the inside dimensions; or
3. Given an outside and an inside dimension, add the dimensions and subtract one metal thickness for each 90° bend.

Quarter Circle

1. Given the outside diameter, subtract the thickness from the diameter and then calculate one-quarter of a circumference; or
2. Given the inside diameter, add the thickness to the diameter and then calculate one-quarter of a circumference.

Circle

1. Given the outside diameter, subtract the thickness from the diameter and then calculate the circumference; or
2. Given the inside diameter, add the thickness to the diameter and then calculate the circumference.

Inches Converted to Decimals of Feet

Inches	Decimal of a Foot	Inches	Decimal of a Foot	Inches	Decimal of a Foot
⅛	.01042	3⅛	.26042	6¼	.52083
¼	.02083	3¼	.27083	6½	.54167
⅜	.03125	3⅜	.28125	6¾	.56250
½	.04167	3½	.29167	7	.58333
⅝	.05208	3⅝	.30208	7¼	.60417
¾	.06250	3¾	.31250	7½	.62500
⅞	.07291	3⅞	.32292	7¾	.64583
1	.08333	4	.33333	8	.66666
1⅛	.09375	4⅛	.34375	8¼	.68750
1¼	.10417	4¼	.35417	8½	.70833
1⅜	.11458	4⅜	.36458	8¾	.72917
1½	.12500	4½	.37500	9	.75000
1⅝	.13542	4⅝	.38542	9¼	.77083
1¾	.14583	4¾	.39583	9½	.79167
1⅞	.15625	4⅞	.40625	9¾	.81250
2	.16666	5	.41667	10	.83333
2⅛	.17708	5⅛	.42708	10¼	.85417
2¼	.18750	5¼	.43750	10½	.87500
2⅜	.19792	5⅜	.44792	10¾	.89583
2½	.20833	5½	.45833	11	.91667
2⅝	.21875	5⅝	.46875	11¼	.93750
2¾	.22917	5¾	.47917	11½	.95833
2⅞	.23959	5⅞	.48958	11¾	.97917
3	.25000	6	.50000	12	1.00000

Copyright by The Goodheart-Willcox Co., Inc.

Summary of Formulas

Perimeter of a Square

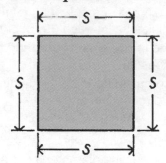

P = Perimeter
S = Side
P = S + S + S + S

Area of a Rectangle

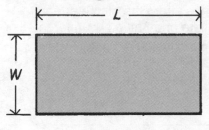

A = Area
L = Length
W = Width
A = L × W

Area of a Square

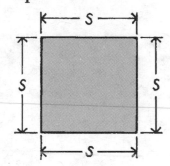

A = Area
S = Side
A = S × S

Perimeter of a Parallelogram

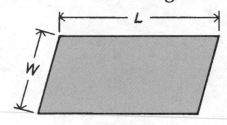

P = Perimeter
L = Length
W = Width
P = L + L + W + W

Perimeter of a Rectangle

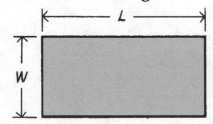

P = Perimeter
L = Length
W = Width
P = L + L + W + W

Area of a Parallelogram

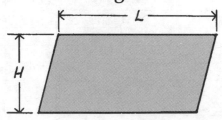

A = Area
L = Length
H = Height
A = L × H

Copyright by The Goodheart-Willcox Co., Inc.

Perimeter of a Trapezoid

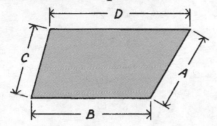

P = Perimeter

A = Length of a Side

B = Length of a Side

C = Length of a Side

D = Length of a Side

P = A + B + C + D

Perimeter of a Triangle

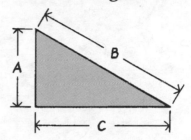

P = Perimeter

A = Length of a Side

B = Length of a Side

C = Length of a Side

P = A + B + C

Area of a Trapezoid

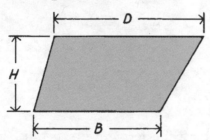

A = Area

B = Length of a Parallel Side

D = Length of a Parallel Side

H = Height

$$A = \frac{1}{2} \times (B + D) \times H$$

Note: To use this formula, first add the two figures inside the brackets, then perform the multiplication.

Area of a Triangle

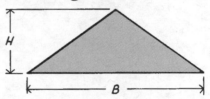

A = Area

B = Length of the Base

H = Height

$$A = \frac{1}{2} \times B \times H$$

Copyright by The Goodheart-Willcox Co., Inc.

Circumference of a Circle

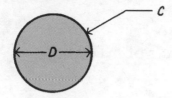

C = Circumference
D = Diameter
π = 3.14
$$C = \pi \times D$$

Perimeter of a Half Circle

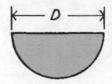

P = Perimeter
D = Diameter
π = 3.14
$$P = \frac{\pi \times D}{2} + D$$

Note: To use this formula, first perform the calculations inside the brackets, then add the result to D.

Diameter of a Circle

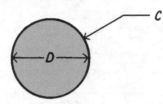

D = Diameter
C = Circumference
π = 3.14
$$D = \frac{C}{\pi}$$

Area of a Half Circle

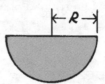

A = Area
R = Radius
π = 3.14
$$A = \frac{\pi \times R \times R}{2}$$

Note: To use this formula, first perform the multiplication, then divide the result by 2.

Area of a Circle

A = Area
R = Radius
π = 3.14
$$A = \pi \times R \times R$$

Area of the Curved Surface of a Cylinder

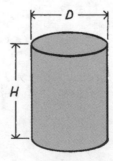

A = Area
D = Diameter
H = Height
π = 3.14
$$A = \pi \times D \times H$$

Copyright by The Goodheart-Willcox Co., Inc.

Perimeter of a Semicircular Sided Shape

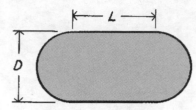

P = Perimeter

D = Diameter

L = Length of Straight Side

$$P = (\pi \times D) + (2 \times L)$$

Note: To use this formula, first perform the multiplication inside the brackets, then add the two results together.

Area of a Semicircular Sided Shape

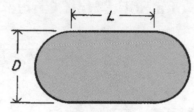

A = Area

D = Diameter

L = Length of Straight Side

$$A = \frac{\pi \times D \times D}{4} + (L \times D)$$

Note: To use this formula, first calculate each of the parts in brackets, then add the two results together.

Volume of a Cube

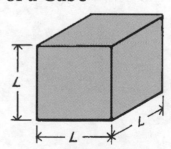

V = Volume

L = Length of a Side

$$V = L \times L \times L$$

Volume of a Rectangular Solid

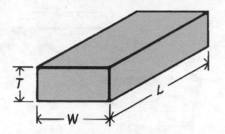

V = Volume

T = Thickness

W = Width

L = Length

$$V = T \times W \times L$$

Volume of a Solid Parallelogram

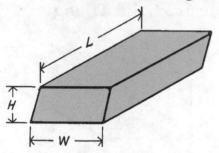

V = Volume

H = Height

W = Width

L = Length

$$V = H \times W \times L$$

Copyright by The Goodheart-Willcox Co., Inc.

Volume of a Solid Trapezoid

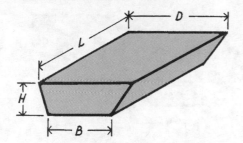

V = Volume
H = Height
B = Length of a Parallel Side
D = Length of a Parallel Side
$$V = \frac{1}{2} \times (B + D) \times H \times L$$

Note: To use this formula, first add the two figures inside the brackets, then perform the multiplication.

Volume of a Cylinder

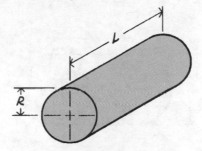

V = Volume
R = Radius
L = Length
$\pi = 3.14$
$$V = \pi \times R \times R \times L$$

Volume of a Solid Triangle

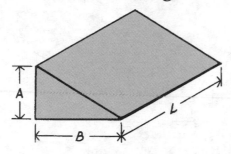

V = Volume
A = Length of a Side
B = Length of a Side
L = Length of a Side
$$V = \frac{1}{2} \times A \times B \times L$$

Volume of a Solid Half Circle

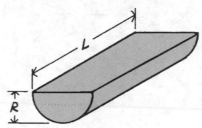

V = Volume
R = Radius
L = Length
$\pi = 3.14$
$$V = \frac{\pi \times R \times R \times L}{2}$$

Note: To use this formula, first perform the multiplication, then divide the result by 2.

Copyright by The Goodheart-Willcox Co., Inc.

Volume of a Solid Semicircular Sided Shape

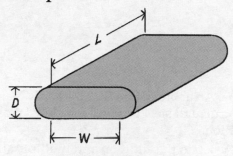

V = Volume

D = Diameter

W = Length of Straight Side

L = Length

π = 3.14

$$V = \frac{\pi \times D \times D \times L}{4} + (D \times W \times L)$$

Note: To use this formula, first perform the calculations inside the brackets, then add the two results together.

Copyright by The Goodheart-Willcox Co., Inc.

Glossary

A

acetylene gas: A gas which is used in combination with oxygen as a fuel for welding and cutting.

addition: The process of combining two or more individual numbers to form a single number that usually has a higher value than any of the previously individual numbers.

alloy: An *alloy* is a blend of two or more elements, usually metal. Alloys are made to achieve certain special characteristics which do not occur in pure metals. See *stainless steel*.

aluminum flux: A flux designed for brazing aluminum. See *flux*.

angle: The opening between two intersecting lines.

angular measure: Refers to measuring the angle formed by two intersecting lines.

approximate numbers: Numbers that are not perfectly accurate. For example, the number 16,297 can be written as the approximate number 16,300. See *rounding*.

Arabic number system: The number system developed in the Arab culture and is now in common usage; made up of the ten digits 0, 1, 2, 3, 4, 5, 6, 7, 8, 9. See *decimal number system*.

area: The entire surface measure of a shape.

ASTM (American Society for Testing of Materials): Organization that performs a variety of testing services and set standards for industry. For example, in welding they have set up certain requirements and established various classifications for welding electrodes.

B

basic operations: The mathematical operations of addition, subtraction, multiplication, and division.

borrow: A step in subtraction and division processes where a number from the place value to the left is used to create a new number. See *division* and *subtraction*.

brass: An alloy made up of copper and zinc.

C

C shape: An abbreviation for Standard Channel. A structural steel product with a "[" cross section.

carry: A step in the addition and multiplication processes of whole numbers in which the calculated value of a number combination is more than one single digit and the higher valued digits are added to the calculated value of the numbers in the column to the left. See *addition* and *multiplication*.

cast iron: Cast iron is an alloy of iron and carbon. To be classified as cast iron, the carbon content must be more than 1.7%. See *steel*.

centimeter (cm): A metric measurement of length. It is equal to 10 millimeter, 0.01 meter, or approximately 2.54".

circular measure: Refers to measuring curved lines.

circumference: The *perimeter* of a circle.

cold galvanizing compound: A paste product that can be applied to steel for the purpose of protecting it from corrosion. See *galvanizing*.

cold rolled steel: A process used in making steel. Cold rolled steel is produced in its final form by passing steel between successive rollers. Both the steel and the rollers are cold, which results in a very accurate steel product with a smooth, shiny surface. Compare with *hot rolled steel*.

common denominator: When two or more fractions have the same denominator, they are said to have *common denominators*. For example, both ³⁄₁₅ and ²⁄₁₅ have a denominator of 15.

common fraction: A fractional part expressed with a numerator and a denominator. Compare with *decimal fraction*.

complex fraction: A fraction which has a fraction for the numerator and a fraction for the denominator.

Copyright by The Goodheart-Willcox Co., Inc.

D

decimal fraction: A number which is less than a whole number and is expressed with decimal. For example, 0.5. Compare with *common fraction*.

decimal number system: The number system in common usage having the ten digits 0, 1, 2, 3, 4, 5, 6, 7, 8, 9. See *Arabic number system*.

decimal point: The point (.) in a decimal number which divides the whole part of the number from the fractional part of the number. For example, 12.75.

degree: A unit of measure used for angles; $\frac{1}{360}$ of a circle.

denominate numbers: Numbers that represent measurements. For example, 17 feet is a denominate number; but the number 17 by itself is not a denominate number.

denominator: The bottom number in a common fraction.

diameter: The straight line distance across a circle and passing through its center.

difference: The answer that results when one number is subtracted from another. See *remainder*.

dividend: In a division operation, it is the number being divided.

division: The process of finding how many times one number (the divisor) can be "contained" in another number (the dividend).

divisor: In a division operation, it is the number that is doing the dividing. The dividend is divided by the divisor.

double extra strong pipe: Pipe is generally classified as Standard, Extra Strong, and Double Extra Strong. The main difference among these is the wall thickness. For example, a Standard 2" pipe has a wall thickness of 0.154" while a 2" Double Extra Strong pipe has a wall thickness of 0.436".

E

equilateral triangle: A triangle in which all sides are the same length and all angles are equal (60°).

equivalent fraction: Fractions that are equal in value to each other but have different numerators and denominators than each other. For example, ½, ²⁄₄, and ³⁄₆ are equivalent fractions. See *higher terms* and *lowest terms*. Equivalency of any two fractions can be checked by reducing each fraction to its lowest terms.

extra strong pipe: Pipe is generally classified as Standard, Extra Strong, and Double Extra Strong, the main difference being the wall thickness. For example, a Standard 2" pipe has a wall thickness of 0.154" while a 2" Extra Strong Pipe has a wall thickness of 0.436".

F

fabricate: To construct by putting parts together.

filler material: The rod that a welder adds to the weld in order to increase the amount of material making up the joint. Also, commonly referred to as a welding rod.

fillet weld: A weld with a triangular cross section.

flat-out: A term describing the original length of a piece of metal that has been permanently bent, shaped, or formed.

flux: A material used to prevent the formation of oxides during the brazing (or soldering) process. It is usually in a paste form and is rubbed on the filler rod and the joint. Its main ingredients are boric acids, chlorides, fluorides, and wetting agents.

G

gage (GA): One of any numbering systems used to identify the various thicknesses of sheet steel.

gallon: A standard American measurement of liquid capacity equal to 231 in³. The Canadian gallon contains 277.42 in³.

galvanizing: This is the process of applying a thin coating of zinc to steel or iron in order to protect it from corrosion. There are a number of methods for doing this, the most common being the hot dip method where the steel is dipped in molten zinc. See *cold galvanizing compound*.

gear pump: A pump for delivering fluids. It causes a flow by passing the fluid between the teeth of two rapidly turning meshed gears.

gram: A metric measurement of weight; $\frac{1}{1000}$ kg.

Copyright by The Goodheart-Willcox Co., Inc.

H

heat exchanger: A device used to transfer heat from a medium (such as water or air) flowing on one side of a barrier to a medium on the other side of the barrier.

higher terms: A common fraction expressed as an equivalent fraction having higher values for numerator and denominator that could be reduced. See *equivalent fraction*.

hot rolled steel: This refers to a manufacturing process used in making this steel. While the steel is still quite hot from manufacturing, it is passed through rollers to the required shape. Because it is being rolled while hot, this steel is rather dark, and is not as accurately formed as cold rolled steel.

hypotenuse: The sloping line of a right triangle. It is the longest of the three sides.

I

improper fraction: A common fraction with a numerator larger than the denominator. For example, $7/6$.

inside corner: The side of a permanently bent piece of metal which is compressed and forms a concave shape.

invert the divisor: The process in the division of fractions whereby the divisor is turned upside down thus converting the division to a multiplication.

isosceles triangle: A triangle in which two of the three sides are of equal length and two of the three angles are equal.

K

kilogram (kg): A metric measurement of weight. It is equal to one thousand grams or approximately 2.205 lb.

L

L-shape: An abbreviation for angle. A structural steel shape with a "L" cross section.

LCD. See *lowest common denominator*.

linear measure: Refers to measuring the straight line distance between two points.

liter: A metric measurement of capacity. It is a little more than a quart (1.06 qt.)

lock washer: A washer designed in such a way that it tends to enhance the holding power of a nut and bolt.

lower terms: A common fraction having numerator and denominator reduced to lesser values than those of equivalent fractions.

lowest common denominator (LCD): A denominator for two or more fractions which is common to all the fractions and which is the lowest number possible.

lowest terms: A common fraction expressed as an equivalent fraction having the lowest possible values of numerator and denominator. See *equivalent fraction*.

M

mass: A measure of the amount of material in an object.

meter (m): A metric measurement of length. It is equal to 1000 mm (or 100 centimeters). It is about 3″ longer than a yard.

metric system: A simple and accurate system of measuring all things that are measurable. Also known as SI Metrics or SI (Systeme Internationale).

MIG: A commonly used colloquial term referring to a welding process that uses a continuous consumable wire electrode that is fed through the torch. The weld itself is covered and protected by a shield of gas which is also fed through the torch. The proper name for the MIG process is *gas metal arc welding (GMAW)*.

MIG welder: This refers to the torch, wire, feed, power, gas flow, etc., that make up the equipment necessary for MIG welding (GMAW). The equipment can be completely automatic or semiautomatic.

millimeter (mm): The smallest metric measurement of length. It is $1/1000$ of a meter or approximately 25.4″.

minuend: In a subtraction operation, it is the largest of the two numbers. The subtrahend is subtracted from the minuend See *subtrahend*.

minute: A unit of measure used for angles. $1/60$ of a degree. See *degree*.

mixed number: A number consisting of a whole number and a fraction. For example, $5\frac{3}{8}$.

multiplicand: In a multiplication operation, this is the number that is being multiplied. The multiplicand is multiplied by the multiplier, or the multiplicand is being added to itself as many times as the value of the multiplier.

Copyright by The Goodheart-Willcox Co., Inc.

multiplication: A fast and simplified method of adding that involves the repeated addition of a single number (the multiplicand) a set number of times (the multiplier).

multiplier: In a multiplication operation, it is the number doing the multiplying. The multiplicand is added to itself the number of times equivalent to the value of the multiplier.

N

numerator: The top number in a common fraction.

O

of: A word used to express the multiplication of fractions. For example, ⅕ of ¾ means ⅕ × ¾.

outside corner: The side of a permanently bent piece of metal which is stretched and forms a convex shape.

oxyacetylene: A combination of oxygen and acetylene gas which, because it burns at an intense heat, is used as a fuel in welding and cutting. See *acetylene gas*.

P

parallelogram: A shape having four sides. The sides opposite each other are the same length and are parallel. The angles opposite each other are equal. These are often described as looking like rectangles that have been tilted.

payload: The maximum weight that a carrier (railroad car, airplane, etc.) can transport.

percent (%): Parts of the whole of anything when it is divided into 100 equally sized parts.

perimeter: The distance around a shape.

pi (π): A Greek letter pronounced *pie*. It represents the number 3.14159...

place value: The value of a number according to its place in a line of numbers. For example, the number 497 has three digits, each with a specific place value. Since the 7 is in the units digit, it has a value of 7. Since the 9 is in the tens digit, it has a value of 90. Since the 4 is in the hundreds digit, it has a value of 400. Each of these numbers in their place value combines to form the number 497.

product: The final calculated value of a multiplication operation.

proper fraction: A common fraction with a numerator smaller than the denominator. For example, ⅔.

protractor: An instrument used to measure angles.

Q

quotient: It is the answer arrived at in a division operation. It is the number of times the divisor fits into the dividend.

R

radius: The straight line distance from the center of a circle to its edge.

rectangle: A shape having four sides. Similar to a square except that two sides are equally longer or shorter. See *square*.

reducing: The process of changing improper fractions to mixed or whole numbers. Also, the process of expressing a common fraction in its lowest terms.

remainder: The amount remaining when a divisor will not divide evenly into a dividend. See *division*. It may also refer to the answer that results when one number is subtracted from another. See *difference* and *subtraction*.

right triangle: A triangle in which one angle is 90°.

root opening: This refers to the gap that may (or may not) exist between two pieces to be welded. If the two pieces touch each other, the root opening is zero.

rounding: The process of modifying accurate numbers into approximate numbers. After determining to which place value to round, evaluate the digit immediately to the right of that place value to determine whether to round up or down. If the value is 5 or higher, round upwards by 1. If below 5, leave the determined place value as is. Next, reduce down to zero all the digits to the right of the approximated digit. For example, in rounding the number 115,389 to the hundreds place, look to the number in the tens digit. Since it is an 8, round the number in the hundreds digit upward by 1. Zero out the digits to the right of the hundreds digit. The 115,389 has been rounded to the approximate number 115,400.

rounding the decimal: The process of writing accurate or unending decimal numbers as approximate decimal numbers. For example, 15.69032 can be rounded to 15.69. See *rounding*.

Copyright by The Goodheart-Willcox Co., Inc.

S

S shape: An abbreviation for Standard Shape. A structural steel shape with a slope on the inside faces of the flanges of 1:6.

screw machine: A high-production, automatic machine originally designed to produce screws and other threaded fasteners. However, the name "screw machine" is highly misleading since they are used to form a large variety of parts including items such as welding tips.

second: A unit of measure used for angles: $\frac{1}{60}$ of a minute. See *minute*.

SI Metric System: The metric system started about 200 years ago in France. As time passed, new versions and variations were randomly added to the system, causing unnecessary complications. Something had to be done, so in 1960, after lengthy international discussions, the International System of Units was established. The system is referred to as SI. It is the official, modernized metric system that is now replacing all former systems of measurement, including former versions of the metric system.

silo: A tall, cylindrical structure in which grain is stored.

solder: This is the filler metal used in soldering, which is a joining method in which the weldment is heated but not to the point of melting. The solder melts below 800°F and joins the parts together. Solder is usually an alloy or combination of tin and lead.

square: A shape having four equal length sides. The opposite sides are parallel, all sides are the same length, and all angles in the square are 90°.

square units: Used in area measurement. Common units are square feet, square inches, and square millimeters.

stainless steel: An alloy steel that has a strong resistance to corrosion because of the addition of chromium as an alloy. See *Alloy*.

steel: Steel is an alloy of iron and carbon. To be classified as steel, the carbon content must be 1.7% or less. If it is more than 1.7%, the metal is classified as cast iron. Other ingredients may be added to the steel for special purposes. See *stainless steel*. Steel that contains no other ingredients besides iron and carbon is called plain carbon steel.

Steel Construction Manual: A book produced by the American Institute of Steel Construction, providing a full description of structural shapes. The Canadian equivalent is produced by the Canadian Institute of Steel Construction.

subtraction: A process of "taking away" that determines how much larger one number is than another.

subtrahend: In a subtraction operation, it is the number being subtracted. The subtrahend is subtracted from the minuend. See *minuend*.

sum: The name of the final calculated number in an addition problem.

supplementary units: The radian and steradian are the two supplementary forms of measurement which are not included in the seven basic units of SI. They are used for measuring angles in two dimensions and three dimensions.

surface grinder: A precision grinding machine on a pedestal. The workpiece is clamped in place on a movable work table and then passed under a revolving grinding wheel.

surfacing: A welding process whereby weld material is deposited on a surface in order to build it up. For example, the worn surface of a bulldozer blade may be built up by surfacing and then machined to proper dimension.

T

terms: The numerator and denominator together are the terms of a fraction.

3-4-5 triangle: A triangle with the length of the hypotenuse at 5, one side at 4, and the remaining side at 3, thus producing a right angle.

time card: A record of the hours an employee works on a day-to-day basis.

tolerance (±): A specific range given of a dimension that a tradesperson may work within and still produce the desired results.

trapezoid: A shape having four sides. Two of the sides are parallel, but the other two sides are not parallel to each other.

triangle: A shape having three straight sides.

Copyright by The Goodheart-Willcox Co., Inc.

U

US Customary system: A common measuring system using feet and inches. For example, twelve inches make up one foot. Occasionally called *conventional system*.

US standard gage: One of the most popular gages used in classifying sheet steel thicknesses.

V

V-groove joint: A type of joint where two flat pieces are beveled on their sides to form a V-shape.

vice versa: The order of something that has been changed from that of a previous statement. For example, A trusts B and vice versa; B trusts A.

volume: The amount of space an object occupies.

W

W shape: An abbreviation for *wide flange shape*. A steel shape characterized by a constant thickness of the flange. In general, it has a greater flange width and a relatively thinner web than S shapes.

weight: The gravitational force exerted by the earth (or another celestial body) on an object.

weldment: Any metal object made primarily by welding.

Copyright by The Goodheart-Willcox Co., Inc.

Index

Copyright by The Goodheart-Willcox Co., Inc.

Copyright by The Goodheart-Willcox Co., Inc.

Answers to Odd-Numbered Practice Problems

Section 1: Whole Numbers

Unit 1: Introduction to Whole Numbers, page 13

1.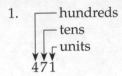
   hundreds
   tens
   units
   471

3.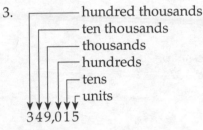
   hundred thousands
   ten thousands
   thousands
   hundreds
   tens
   units
   349,015

5. 694,700
7. 17,200
9. 29,900
11. 946,000
13. 11,000
15. 449,000
17. 1,295,000
19. 13,001,000
21. A. 250 psi, B. 11,500 ft², C. five acres, D. 17.5 tons, G. $7.65, H. $7.65 per hour, I. 40" per minute, K. 12 dozen

Unit 2: Addition of Whole Numbers, page 19

1. 706
3. 34,472
5. 89,454
7. 37
9. 825,833
11. 1,140
13. 14,000
15. 4,333
17. 1,141,120
19. 88"

21. 366,433 welding rods
23. 159 hours
25. 1,766 women
27. 541"
29. 552"
31. 209"
33. 9,028 hours

Unit 3: Subtraction of Whole Numbers, page 29

1. 664
3. 1,746
5. 20,019
7. 42,630
9. 1,015,423
11. 5,722
13. 7,222
15. 27
17. 93,640
19. 805 brackets
21. 34,730 lb of 16 gage
23. 20,831 lb
25. 1,980 of #51 tips
27. 413 lb
29. 32,835 lb of steel
31. 191"
33. 91"
35. 158" of channel
37. 406" of flat bar

Unit 4: Multiplication of Whole Numbers, page 41

1. 408
3. 3,920
5. 174,915
7. 4,748,576
9. 1,876,070

Copyright by The Goodheart-Willcox Co., Inc.

11. 4,037,688

13. 2,508,156

15. 4,680

17. 637,512

19. 18,914 spot welds

21. 1,053 studs

23. 52,254 holes drilled

25. 7,948 lb

27. 63,138′ of 1½′ pipe

29. 238,954′ of pipe

31. 42 lb

33. 26,588″

35. 765 lb

37. 3,500 lb

Unit 5: Division of Whole Numbers, page 53

1. 27

3. 103

5. 9,007 r 1

7. 7

9. 400

11. 25 r 40

13. 64 r 44

15. 25 r 70

17. 19 r 4

19. 61 lb

21. 6 pieces

23. 383,916 lb

25. 183′

27. 8″

29. 6″

31. 985 pieces

33. 221 studs

Section 2: Common Fractions

Unit 6: Introduction to Common Fractions, page 61

1. $^{10}/_{16}$

3. $^{98}/_{98}$

5. $^{7}/_{8}$

7. $^{55}/_{65}$

9. $^{80}/_{100}$

11. yes

13. no

15. yes

17. yes

19. 4½

21. 17⅝

23. 3½

25. 10$^{1}/_{30}$

27. 2

29. $\dfrac{51}{4}$

31. $\dfrac{233}{8}$

33. $\dfrac{1670}{9}$

35. $\dfrac{110}{10}$ or reduced to $^{11}/_{1}$

37. ½

39. ⅞

41. $^{4}/_{41}$

43. $^{3}/_{5}$

45. $^{101}/_{200}$

Unit 7: Addition of Fractions, page 67

1. $^{11}/_{18}$

3. $^{168}/_{219}$

5. 2 $^{23}/_{50}$

7. 2 $^{23}/_{60}$

9. 1 $^{11}/_{24}$

11. 1 $^{2,451}/_{16,192}$

13. 23 $^{7}/_{16}$

15. 158 $^{97}/_{204}$

17. $^{53}/_{238}$

19. 402¼ hours

21. 65 $^{11}/_{16}$″

23. 22 $^{9}/_{16}$″

25. 41 $^{61}/_{64}$″

27. 79 $^{45}/_{64}$″

Copyright by The Goodheart-Willcox Co., Inc.

29. 92 $^{15}/_{32}$″
31. 73 $^{7}/_{32}$″
33. 505 $^{55}/_{64}$″

Unit 8: Subtraction of Fractions, page 75

1. $^{1}/_{3}$
3. $^{31}/_{101}$
5. $^{1}/_{16}$
7. 689$^{15}/_{16}$
9. $^{111}/_{238}$
11. 15$^{1}/_{3}$
13. $^{2}/_{3}$
15. $^{7}/_{15}$
17. 12$^{2}/_{3}$
19. 5′-2$^{13}/_{16}$″
21. 82$^{3}/_{8}$ miles
23. 72$^{1}/_{8}$ miles
25. 2$^{7}/_{8}$″
27. 66$^{45}/_{64}$″
29. 9$^{5}/_{16}$″
31. 7$^{9}/_{16}$″
33. Length B: 5$^{5}/_{8}$″
35. Dimension A: 19$^{13}/_{32}$″
37. 30$^{15}/_{16}$″
39. 10$^{11}/_{32}$″
41. 5$^{7}/_{32}$″

Unit 9: Multiplication of Fractions, page 87

1. $^{1}/_{15}$
3. $^{60}/_{671}$
5. $^{13}/_{72}$
7. $^{3}/_{4}$
9. 39$^{3}/_{20}$
11. 87$^{57}/_{64}$
13. $^{20}/_{63}$
15. 25$^{11}/_{64}$
17. 2$^{5}/_{8}$
19. 3$^{11}/_{16}$ miles
21. 200$^{1}/_{4}$ hours
23. 71$^{1}/_{2}$′

25. 10,802,500 bolts
27. 237$^{7}/_{16}$″
29. 1,125$^{3}/_{16}$ lb
31. 270 rods
33. 3,319$^{1}/_{32}$ lb
35. 378 lb
37. 105,252″
39. 68,316$^{3}/_{4}$″ or 5,693′ $^{3}/_{4}$″

Unit 10: Division of Fractions, page 95

1. 1$^{1}/_{15}$
3. $^{20}/_{21}$
5. $^{9}/_{16}$
7. 25$^{1}/_{5}$
9. 21$^{5}/_{7}$
11. 2$^{33}/_{57}$
13. $^{29}/_{36}$
15. 1$^{9}/_{112}$
17. $^{5}/_{12}$
19. 8 pieces
21. 35 miles per hour
23. 49 pieces
25. 236$^{13}/_{15}$ ft^3
27. 17$^{5}/_{8}$″
29. 10 pieces
31. 13$^{19}/_{64}$″

Section 3: Decimal Fractions

Unit 11: Introduction to Decimal Fractions, page 105

1. 100.01
3. 14.00125
5. 0.707
7. 0.866
9. $^{9}/_{20}$
11. $^{267}/_{2,000}$
13. $^{3,333}/_{10,000}$
15. 0.5
17. 0.03125
19. 0.625

Copyright by The Goodheart-Willcox Co., Inc.

21. 0.9
23. 0.5
25. 0.1
27. 0.96
29. 0.09
31. 0.91
33. 0.314
35. 0.009
37. 0.443
39. 0.67
41. 0.11
43. 0.10
45. 0.938
47. 0.063
49. 0.767
51. 0.0156
53. 0.1235
55. 0.9659

Unit 12: Addition and Subtraction of Decimal Fractions, page 111

1. 13.4
3. 363.398
5. 52.66532
7. 222.49
9. 104.3088
11. 0.7
13. 17.6982
15. 4.093
17. 5.0644
19. 345.4892
21. 1.757″
23. 88.328″
25. + $97.55
27. Rod B: 1.413″
29. Rod D: 2.7198″
31. Rod F: 4.0266″
33. Rod H: 5.3334″
35. 30.3588″
37. Block A: 1.45625
39. Block C: 1.24585
41. 15.8724″

Unit 13: Multiplication of Decimal Fractions, page 119

1. 0.3026
3. 70.356
5. 1,001.01001
7. 6.66
9. 1,093.75
11. 29.9832
13. 0.00008
15. 1.13918
17. 195,000
19. 65.20293
21. 1.367631
23. 68,200 lb
25. $22.40
27. 51.75 ft³
29. 83,145.258 lb
31. $240.60
33. 3,504.879″
35. 112.271″
37. 2,038.06528 lb
39. 1,872.54846 lb

Unit 14: Division of Decimal Fractions, page 127

1. 3,500
3. 0.197
5. 8
7. 0.07
9. 1.00
11. 136.96
13. 3.000
15. 0.017
17. 0.719
19. 16 hours
21. 750 ball bearings
23. 4 blocks
25. 623 items
27. 364 squares
29. 54 applications
31. 5.6233″

Copyright by The Goodheart-Willcox Co., Inc.

Section 4: Measurement

Unit 15: Linear Measure, page 137

1. 108 mm, 4¼″
3. 74 mm, 2¹⁵/₁₆″
5. 18 mm, ²³/₃₂″
7–11.   Individual answers will vary.
13. 423″
15. 544.5″
17. 6,042″
19. 63,360″
21. 3³/₁₆″
23. 145″
25. ⅛′
27. 7′-11.75″
29. 17′-11¹/₃₂″
31. 65′-7.96″
33. 6,993′-1″
35. 1¹/₃₂′
37. 304.8 mm
39. 254.0 mm
41. 0.4 mm
43. 15.9 mm
45. 1605.0 mm
47. 209.6 mm
49. 39.370″
51. 0.039″
53. 2.579″
55. 466.535″
57. 7.677″
59. 207.874″
61. 15¾″
63. 48″
65. 19⅜″
67. 19¹⁴/₁₆″
69. 1″
71. 21¹/₁₆″
73. 11²¹/₃₂″
75. 730⁶/₃₂″
77. ³¹/₃₂″
79. 68²/₆₄″
81. 12³⁵/₆₄″
83. 632²⁶/₆₄″

85. 249 mm; 247 mm
87. 1.345″; 1.335″
89. 15 mm; 14 mm
91. 20.1 mm; 19.7 mm
93. 3⁹⁰/₁₀₀″; 3⁹³/₁₀₀″; 3⁸⁷/₁₀₀″
95. 87⁶⁰/₁₀₀″; 87⁶³/₁₀₀″; 87⁵⁷/₁₀₀″
97. ⁷/₁₆″; ⁵/₁₆″
99. 8¹³/₁₆″; 8¹¹/₁₆″
101. 15⁵/₁₆″; 15³/₁₆″
103. 0.438″; 11.13 mm
105. 0.500″; 12.70 mm
107. 0.531″; 13.49 mm
109. 4.969″; 126.21 mm
111. 37.5 mm; 37.0 mm
113. 149.5 mm; 149.0 mm
115. 160.1 mm; 159.6 mm
117. 5 mm; 0.197″; ¹³/₆₄″
119. 78 mm; 3.071″; 3⁵/₆₄″
121. 13 mm; 0.512″; ³³/₆₄″
123. 50 mm; 1.969″; 1⁶²/₆₄″

Unit 16: Angular Measure, page 151

1. 90°
3. 20°
5. 215°
7. 10° 38′
9. 75° 21′
11. 200° 39′ 23″
13. 66° 50′ 55″
15. 28° 2′ 20″
17. 33° 59′ 2″
19. 293°43′ 42″
21. 11° 52′ 28″
23. 283° 14′ 56″
25. 909° 50′ 12″
27. 1,239° 16′ 42″
29. 31° 14′ 9″
31. 30°
33. 51° 25′ 43″
35. 8° 27′ 59″
37. 90°; 85°
39. 40° 35′; 33° 25′
41. 30° 30″; 29° 59′ 30″

Copyright by The Goodheart-Willcox Co., Inc.

43. 212° 47′; 212° 17′

45.

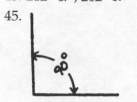

47.

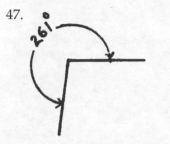

49.

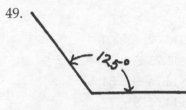

51.

53. 30°

55. 20°

Unit 17: Four-Sided Figure Measure, page 163

1. 5,040 in²
3. 1,400 in²
5. 65 in²
7. 1 ft²
9. 111 ft²
11. 694⁴⁄₁₀ ft²
13. 58064 mm²
15. 88468 mm²
17. 645 mm²
19. 2.4 in²
21. 0.2 in²
23. 1.0 in²
25. 65,985 in²
27. 252537 mm²
29. 9,919,872 in²
31. 2,351.25 in²
33. 39354720 mm²

35. 13222 mm²
37. 347½″
39. 6,217½ in²

Unit 18: Triangular Measure, page 173

1. 12000 mm
3. 161′
5. 336″
7. 698.705 in²
9. 5565174.8 mm²
11. 2,812.5 ft²
13. 82″
15. 4,896 in²
17. 332.544′
19. 556.6′
21. 23500 mm

Unit 19: Circular Measure, page 187

1. 2,582 in²
3. 166′-5″
5. 12¹⁵⁄₁₆ in²
7. 2,499 in²
9. 4,068 in²
11. 576334 mm²
13. 395.3″
15. 156000 mm
17. 450″

Section 5: Volume, Weight, and Bending Metal

Unit 20: Volume Measure, page 201

1. 1,443.75 in³
3. 11,968 gallons
5. 127 gallons
7. 79 ft³
9. 13,824 in³
11. 737415 mm³
13. 269.28 gallons
15. 155 in³
17. 176839.27 mm³
19. 11¾ yd³

Copyright by The Goodheart-Willcox Co., Inc.

21. 8,117.34 in³
23. 1,200,000 mm³
25. 15,660 in³
27. 4 ft³
29. 281893.5 mm³
31. The rectangular container
33. 926¼ in³
35. 42,390 in³
37. 603,750 mm³

Unit 21: Weight Measure, page 217

1. 15 kg
3. 949 lb
5. 214.5 kg
7. 370 lb
9. 16.73 lb
11. 1,706.4 lb
13. 7,706 lb
15. 227 lb
17. 10635 grams
19. 181.3 lb

Unit 22: Bending Metal, page 235

1. 38'-3" (or 459")
3. 1068 mm
5. 556 mm
7. 78½"
9. 208$\frac{1}{32}$"
11. 1832 mm
13. 87¾"
15. 1257 mm

Section 6: Percentages and the Metric System

Unit 23: Percentages, page 247

1. $^{13}/_{100}$
3. $^{1}/_{2000}$
5. 10$^{1}/_{10}$
7. 0.75
9. 0.1414
11. 0.00007
13. 75%

15. 3333⅓%
17. 57$^{17}/_{19}$%
19. 12.5%
21. 1.9%
23. 0.3%
25. 288.05
27. 28.5
29. 0.21012
31. 22.46%
33. 50%
35. 900%
37. 207 parts per hour
39. 73%
41. $411,450,000
43. $514.95, $92.69, $18.02, $20.60
45. $335.20, $50.28, $11.73, $13.41
47. $421.10, $75.80, $14.74, $16.84
49. $417.74, $71.02, $14.62, $16.71

Unit 24: The Metric System, page 261

1. length, mass, time, electrical current, temperature, luminous intensity, substance
3. Meter
5. 100 cm
7. 15.97 m
9. 2985.4 km
11. 0.0595 km
13. 1524 mm
15. 419 mm
17. 91.44 m
19. 3$^{29}/_{64}$"
21. 3,280'-10$^{5}/_{64}$"
23. 32,808'-4$^{50}/_{64}$" (or reduced to 32,808'-4$^{25}/_{32}$")
25. 1000 ml
27. 0.182 L
29. 6700 ml
31. 2.64 gallons
33. 1000 g
35. 39250 g
37. 0.212 g
39. 0.065 kg
41. 14.27 lb
43. 3175 kg
45. 62597 kg

Copyright by The Goodheart-Willcox Co., Inc.